Your Eternal Soul's Journey

Simon S. M. Chin

Printed in 2025 by IngramSpark

Design by Simon Chin

Manufactured in the USA

2025919420

Library of Congress Cataloging-in-Publication Data

Chin, Simon S. M.

Your Eternal Soul's Journey

eBook ISBN: 979-8-9924674-1-3
Paperback ISBN: 979-8-9924674-0-6

Hardcover ISBN: 979-8-9924674-2-0

Table of Contents

Dedication

To my beloved mother

A mother I love wholeheartedly and respect deeply

Preface

This book is for a global audience of mostly young adults and adults. If you want to learn what you can do to fight against the scourge of pandemics, wars, global financial crisis, income inequality, unequal opportunities, overwhelming stress and losing your soul, then this book is for you.

After my beloved mother passed away on July 7, 2023, I decided to write this book. She was an inspiration to me and I want to share the insights I gained throughout my life. I believe it is important to learn from each other's experiences. When I saw my mother's soul leaving her body five minutes after she stopped breathing, it was clear to me that our bodies are just vessels for our souls.

My life has been filled with many blessings. I was lucky to be in the right place at the right time. God gave me a photographic memory and I am grateful. In Kindergarten, I learned fast and skipped first grade. When I left Burma to come to the United States, I was ranked first in my class. Having attended a large high school in San Francisco, I became the head of the student body after I ran for student government for the first time against more experienced and popular students.

God has led the right people into my life at the right time to help me achieve many things. University of California, Berkeley was ranked #1 for Chemistry in the country in 1977, and it was the only university to which I applied. I got in and was given the Alumni Scholarship based on merit. My faculty advisor, Dr. Glenn T. Seaborg, Nobel Laureate in Chemistry and Second Chancellor of UC Berkeley, gave me an opportunity to do research with him at the Lawrence Berkeley National Laboratory starting in my first quarter as a freshman. I was very lucky and grateful.

After graduating from UC Berkeley with a degree in Chemical Engineering, I was hired by the first company that I interviewed with in Silicon Valley. My trajectory there was swift – starting as an engineer, I was promoted to a managerial role overseeing manufacturing within just 9 months, doubling my salary by the end of my first year of employment. Assuming responsibility for the company's largest operation 6 months after the first promotion, we experienced a period of great success and rapid growth. I was blessed to be able to buy my first house less than two and a half years after graduation from UC Berkeley.

I simultaneously pursued an MBA at Santa Clara University while working full-time - juggling employment's demands with graduate studies' rigors. I actually enjoyed the synergy of applying what I learned in school to my work and what I learned at work to my school projects almost on a daily basis.

On my first trip oversea to transfer technologies to Bosch in Germany, Lufthansa Airlines oversold business class and upgraded me to the first class cabin. The technology transfer projects I managed for Du Pont, the eighth-largest company in the US at the time, contributed to uplifting 800 million people out of poverty in China. The first professional company I founded, Iris Biotechnologies Inc., started trading as a public company in 2008.

Iris applied for patents all over the world, and all our patent applications were granted as patents. I have three patents on Artificial Intelligence Systems for Genetic Analysis. I completed a marathon the first time I ran one, raising money for leukemia and lymphoma cancer research. I then founded Iris Wellness Labs to help patients and physicians with complex medical cases by analyzing the whole human genome (3 billion base-pairs), microbiome, and about 900 metabolites.

My goal is to share my more than forty years of experience in high technology, biotechnology, medicine, and artificial intelligence and fifty years studying Christianity, Buddhism, and other religions. I hope to offer clear ideas to empower people to help themselves and make a positive impact.

Most scholars believe that Darwin withheld from publishing his book, On the Origin of Species, for twenty years because he was unsure about his theory. He was forced to publish because he was afraid that Alfred Russel Wallace was going to publish a competing book first. If Darwin knew what we know now, would he have published his book? Some leading scientists believe that mutations reflect adaptation, not the creation of new animals or plants.

Many people today are dealing with multiple, unprecedented crises at the same time. Some people are trying their best, while others are using unhealthy ways to cope, like drugs and alcohol. We live in challenging times. I want to give back, inspire hope and encourage people to work towards a better future. My intent is to provide evidence of what the truth tells about what has been going on in the world.

I am very grateful for the amazing opportunities I've had in life. Many people have shared their wisdom and showed me love and kindness. I am grateful to all my mentors. My life's journey also had many hardships and injustices that made me stronger and prepared me for the future. As a

Christian who was raised in a Buddhist family and lived in the United States, Europe and Asia, my unique experience helped me to understand and appreciate many different cultures and viewpoints. The United States became the most innovative and successful nation in human history by attracting some of the best minds from around the world.

God is real and I want to share my personal experience, beyond reasonable doubt, about him. God gave you a soul and it is the very essence of who you are. This is your eternal soul's journey in this world. I hope this truth will set you free. God gave us the freedom to choose, and we must choose wisely.

I am grateful to Dr. Wing Hsieh, Mr. James Moshofsky, my sisters Grace Osborne and Catherine Cheah, my nephews Joshua and Daniel Osborne, and others for helping me to review different parts of the book and publish this work.

Introduction

This book aims to share the truth. Your eternal soul is the REAL YOU in totality. You time on Earth is just a blink of an eye in cosmic time. You soul is an eternal luminous essence and can go anywhere in the universe without any time lapse. As a scientist, engineer and entrepreneur, I value evidence. I saw my mother's soul leave her body through the right side of her mouth 5 minutes after she stopped breathing.

My experiences draw from my over 16 years as an engineer and manager in semiconductor technology and 25 years as an entrepreneur in biotechnology, artificial intelligence, and precision medicine to make life better for people. After my beloved mother passed away in 2023, I felt a strong need to share my story and insights.

In 1999, I founded Iris Biotechnologies to improve breast cancer diagnosis and treatment. Iris started trading as a public company in 2008. In 2014, I founded a subsidiary, Iris Wellness Labs, to provide in-depth scientific analysis to give insights to patients and doctors facing complex medical conditions. Sharing my experiences and perspectives in this book can uplift and inspire others.

Caring for my mom during her last three years was an honor for me. Her passing left an immense void in my life, but it also made me reflect deeply on my unusual journey from childhood in Burma to becoming an entrepreneur in Silicon Valley in the United States. I also reflected on studying different religions and history for 50 years. God has always been with me, even when I did not realize it.

We live in a Dopamine Culture of clickbaits on reels of short videos on TikTok, YouTube, and other platforms, scrolling and sending short texts on iPhone, Android, and other smart phones and tablets. We also swipe on various apps and gamble with ease on our choice of communication devices. This culture is a far cry from reading books and newspapers, writing letters, and having real and lasting interpersonal relationships such as long-term friendships and marriage. Dopamine is pleasurable and addictive. Having dopamine easily all the time is not good for us physically, mentally, or spiritually.

Charles Darwin's book "On the Origin of Species" has changed the world for 175 years. I have diligently studied his theory and had discussions with people, including leading scientists in biology and chemistry, for decades. I patiently searched and waited for proof of Darwin's Theory, and I have come to the clear conclusion that life evolving from non-life by chance is impossible. We must teach our assumptions truthfully and clearly. Did we evolve from rodents, or did God create humans?

I am struck by the many perilous challenges facing humanity today – challenges like the pandemic, climate change, tensions between countries, global debt crisis and disruptions caused by new technologies such as artificial intelligence. In this pivotal moment, I feel called to offer insights into some of life's most profound questions about our origins, our purpose, and our ultimate destiny.

The ability to discern is one of the most important skills you'll ever learn. It involves seeing, recognizing, or apprehending something by sight, another sense, or the intellect. Having the skill to discern will help you to succeed in life. We can learn to discern and then teach our children how to discern, who to trust, who not to trust, and why truth matters. Children are our future and we must help them to succeed.

My study of prophecies from the Bible and my analysis of trends in technology, politics, and society have led me to believe we are heading toward the end times foretold in the Book of Revelation. I don't know the precise timing, but I believe humanity stands at a critical spiritual crossroads. The choices we make collectively will shape the fate of our species. This is the last call.

At age 66, my primary allegiance is to truth – the truth of God's love, the truth of science, the truth of history, and the truth about our present circumstances. My aim is not to prove anything or please everyone but simply to bear witness to what God has revealed in my life. I hope it enriches others' spiritual understanding and personal journeys. I consider this book to be the most important work of my life. In this book, I will discuss topics like:

- The nature and importance of our souls' journeys
- The entity that enabled wars resulting in 100 million deaths
- The story of my blessed and grateful life from a humble beginning
- The need to refocus your mind: a global snapshot
- The fulfillment of end-time prophecies from religious texts

My perspective comes from my own lived experiences, and your perspective may differ from mine. I do not intend to judge but simply to share what I have learned in the hopes that it inspires you to re-examine your priorities and life's purpose in light of the precious gift of life itself.

All humans begin their lives at conception. From a biological standpoint, conception at fertilization marks the beginning of a new genetic individual. At this point, a unique genome is formed, setting the stage for potential human development. There is no debate or controversy about that.

Between conception and birth, there are disagreements as to when the growing cells can be called a human. Abortion ends the development of cells that could result in the birth of a baby. There is also no debate or controversy about that.

Whether people argue life began at birth, at 120 days, or at any other time between conception and birth, DNA already defines a person's genetic identity at conception. That is a fact. People can believe whatever they choose to believe, but that won't change the truth. We live in a time of deceit, and we must be extra vigilant, especially with AI used to create "Untruthful clickbaits" online.

Your body is just a temporary vessel for your soul, which lives on after your body dies. Your spirit is your conscience. It gives us the ability to connect and have an intimate relationship with God.

You have "free will" because of the human spirit, which asks whether what you intend to do is right or wrong. It is the human spirit that gives us a consciousness of self, intellect, emotions, fears, courage, resilience, passions, creativity, and the unique ability to comprehend and understand. We can think, feel, love, design, create, and enjoy music, humor, and art because of the spirit.

Jesus said, "God is spirit, and those who worship him must worship in spirit and truth" (John 4:24). Our spirits have an innate knowledge of right and wrong (Romans 2:14–15). The soul of a person is the courtroom where life decisions are made, and I discuss the soul in depth. Identical twins have exactly the same DNA but different souls and spirits with unique personalities, talents, likes, and dislikes.

Do you think God is real? I had a profound experience, beyond a reasonable doubt, that God is real. It happened when I least expected it: what you believe regarding God and what you think have profound consequences for your life. If God is real and everywhere at the same time for all times, then we are all in God, and he knows all our thoughts and memories. We, on the other hand, don't know all that God knows.

As humans, we have limitations. According to our understanding of biology, human life, in our times, is limited to approximately 120 years. Research has shown that when all of the cells ever created in the human body are multiplied by the average time it takes for cells to reach the end of their lives, you get roughly 120 years. This is called the "Hayflick limit, the maximum number of years a human can expect to live." Leonard Hayflick is a professor of anatomy at the University of California, San Francisco, and formerly a professor of medical microbiology at Stanford University.

Our universe consists of 5% visible matter, 27% dark matter, and 68% dark energy. Hubble Telescope and James Watt Space Telescope (JWST) show that our sun is one of approximately 100-400 billion stars in the Milky Way galaxy. Our Milky Way is one of 200 billion to 2 trillion galaxies in the universe.

It is improbable to think that God created life only on Earth. The human life span, less than 120 years, is just a blink of an eye in cosmic time. Humans cannot experience the fullness of life without Earth or something like Earth. That's why we must do what we can to preserve our world.

Section 1: The Nature and Importance of the Soul

Chapter 1
The Nature of the Soul

Most people believe that every human has a body, mind, and spirit. The body is the visible, physical vessel that houses our being. The mind can be understood as the self, the psyche, or the ego — the seat of our consciousness and identity. The spirit, meanwhile, is the unseen part that connects us to the divine — to God.

Identical twins have exactly the same DNA but different souls and spirits, reflected in their unique personalities, talents, likes, and dislikes. Twins are typically raised by the same parents in the same environment, eating the same food, and sharing many experiences. What distinguishes them is not their biology, but their souls — each bearing a distinct purpose and trajectory. Modern psychology confirms that environmental sameness cannot fully account for behavioral and temperamental differences in twins.

A helpful analogy is to think of your brain as a personal computer and your soul as a removable storage device that holds your personal data. The computer has essential software, and more can be installed over time. When in use, it's like your mind being active. The Internet — facilitating connection and flow — resembles the spirit that enables communication with God and others. When the computer is shut down, the data — the soul — still exists. Artificial Intelligence, no matter how advanced, lacks both a soul and consciousness. While it can simulate intelligence, it does not possess self-awareness or moral agency.

Unlike machines, which lack a heart, the soul intertwines mind and heart. It is the seat of belief, emotion, intention, and memory. The soul is the complete inner activity of a person — the invisible essence that gives rise to identity, experience, and growth.

Imagine the entirety of your being: physical, mental, emotional, spiritual, and soulful. Your body moves through the world. Your brain enables thinking. Your emotional body stimulates feelings such as warmth, peace, or sadness. In the Bible, the heart is portrayed as the spiritual center — the place of will and desire. Your mind determines how you act, adapt, and succeed. Your spirit is your connection to divine truth. Your soul is the director, the ultimate observer, and the essence of your eternal self.

The mind is a tool of the soul — not the master. It creates logic, solves problems, and organizes thought. But the deepest wisdom lies beyond the mind, in the stillness of the heart. There, the soul speaks.

Across religious and philosophical traditions, the soul is seen as the immortal essence of a person. It carries identity, values, and memory. Many ancient cultures, including the Egyptians, Hindus, and Greeks, believed in a soul that survives death and continues its journey in another realm. As both a scientist and engineer, I had a profound personal experience that confirmed this for me. I witnessed my mother's soul leaving her body through the right side of her mouth five minutes after her final breath. That moment gave me unshakable certainty that the soul survives the body.

The human spirit is a powerful force within us — a source of courage, imagination, empathy, and strength. It fuels our dreams and enables connection. The peace of God's presence can only be felt through the spirit.

I had my own unexpected spiritual encounter that reinforced my belief in God. While standing on the shore of the Sea of Galilee — at the place where Jesus appeared to his disciples for the third time — I was filled with divine energy. I felt the spirit of God enter me. My ego dissolved. I couldn't perceive any separation between my mind and God. This moment left me with absolute conviction that God is real and has always walked with me.

Every human begins life at conception. According to a 2008 position paper by the American College of Pediatricians, "human life begins at fertilization." This view aligns with embryological science, which states that the zygote formed at fertilization is a genetically unique human organism capable of further development given the right conditions.

God gave you a body, a spirit, and a soul. Your soul was within you even in the womb. This is your soul's journey; your body is a temporary vessel. The soul is the very essence of your being — the part that makes you, you. Your spirit is what allows communion with God.

Your body is the visible part that grows, senses, and expresses. It's the container — the vessel — of the soul. Just like a jewel is protected in a box, the soul is held within the body.

The Bible affirms this truth. Psalm 139:13-16 (NIV) reads:

"For you created my inmost being; you knit me together in my mother's womb. I praise you because I am fearfully and wonderfully made; your works are wonderful, I know that full well…"

The soul is eternal, even as the body ages and fades. It lives on because it is from God, who is infinite and timeless.

Genesis 2:7 tells us, "Then the Lord God formed a man from the dust of the ground and breathed into his nostrils the breath of life, and the man became a living soul." This verse shows that our souls are not metaphorical — they are divine gifts, real and vital.

Jesus echoed this in Matthew 10:28:

"Do not be afraid of those who kill the body but cannot kill the soul. Rather, be afraid of the One who can destroy both soul and body in hell."

This underscores the eternal value of the soul and the importance of nurturing it.

Your beauty — physical, intellectual, or spiritual — can be a blessing or a burden. It is up to you, and God's guidance, to shape what that beauty means. Hopefully, you were born to parents who protected and nurtured you. If you're fortunate, you'll experience deep love and form a family that reflects that love.

Every soul comes into this world with a purpose, a path, and a hunger for meaning. Let us cherish those we love while they are here, for life in this physical realm is fleeting.

Over the past 25 years, my mission has been to create life-saving technologies and innovations to improve human health. But I believe it's time to go beyond physical health. It's time to help people reconnect with their souls and secure their spiritual well-being through God's grace. Our bodies are temporary. Our souls are eternal. They are connected to God and to the universe.

As we continue exploring the nature of the soul, let us remember: spiritual wellness is not secondary to physical health — it is foundational. In the chapters ahead, we'll explore how to care for the soul, understand our divine purpose, and navigate life's challenges through faith, resilience, and revelation.

Chapter 2
Nurturing Your Soul and
Discovering Purpose

Nurturing Your Soul

So, how can we nurture and care for our souls? Nurturing the soul begins with intentional spiritual discipline. One of the most important practices is spending regular time with God. This means setting aside daily moments to pray, reflect, read the Bible, and simply rest in His presence. According to multiple studies, including research from the Barna Group and Pew Research Center, individuals who engage in regular spiritual practices report higher levels of life satisfaction and emotional resilience. When we do this, we feed our souls with the spiritual nourishment they need to grow and thrive. We open ourselves to God's wisdom, guidance, and love, allowing Him to shape us into the people He created us to be.

Another essential way to nurture our souls is to cultivate community with others who support our faith journey. This might mean joining a church group, participating in Bible studies, or spending time with friends who share our values. Proverbs 27:17 says, "As iron sharpens iron, so one person sharpens another." The people around us can have a profound impact on our spiritual growth and overall well-being.

Caring for our souls also involves guarding what we allow into our hearts and minds. Just as we cannot eat junk food and expect to remain physically healthy, we cannot fill our minds with negativity or harmful media and expect our souls to flourish. Philippians 4:8 encourages us: *"Whatever is true, whatever is noble, whatever is right, whatever is pure, whatever is lovely, whatever is admirable—if anything is excellent or praiseworthy—think about such things."* This may mean being selective with the movies, TV shows, music, and books we consume, as well as the thoughts we dwell on daily. Replacing negative self-talk with the truth of God's Word is an intentional act of soul-care.

As we nourish our souls, we begin to bear the *fruit of the Spirit* described in Galatians 5:22-23: *"love, joy, peace, forbearance, kindness, goodness, faithfulness, gentleness and self-control."* These are not just pleasant traits—they are evidence of a vibrant and healthy soul. When these qualities flourish in us, they naturally bless those around us and become a witness of God's transformative power.

Discovering Your God-Given Purpose

One of the most fulfilling aspects of a life devoted to God is discovering the unique purpose and calling He has placed on your life. The Bible tells us in Jeremiah 29:11: *"For I know the plans I have for you,"* declares the Lord, *"plans to prosper you and not to harm you, plans to give you hope and a future."* This verse affirms that God's intentions for each person are rooted in love, hope, and purpose.

A powerful way to begin discovering your purpose is by paying close attention to your passions—the activities that energize you and bring you joy. These joys are often divine hints at the direction your soul is meant to take. Do you love creating art, writing, caring for others, or solving problems? As 1 Peter 4:10 states, "Each of you should use whatever gift you have received to serve others, as faithful stewards of God's grace in its various forms." Your interests are often God's way of equipping you to serve the world in a specific way.

Another way to uncover your purpose is by observing the needs around you and asking God how He might use you to address them. Is there a social issue that breaks your heart? A group of people you feel called to serve? These stirrings often point toward your mission. As theologian Frederick Buechner famously wrote, "Your vocation in life is where your greatest joy meets the world's greatest need."

God often aligns our talents with the needs of others. For example, those drawn to healing may feel called to healthcare or counseling; those with empathy may serve the brokenhearted; those with leadership may help build communities of justice and support. When you ask God to show you His purpose for your life, you may be surprised by the clarity and opportunities that begin to unfold.

It's important to note that discovering your purpose is not a one-time event but an evolving process. It often takes prayer, patience, and faithful steps forward. Sometimes your calling is revealed through challenges, failures, or unexpected detours that later make sense in hindsight.

The Example of Moses

Of course, discovering your purpose is not always easy. There will be seasons of doubt, confusion, and discouragement when the path ahead feels unclear. But remember, God's presence is constant—He is always with you, guiding you, and giving you the strength and wisdom to keep moving forward.

One of the most powerful examples of someone discovering and fulfilling their God-given purpose is the story of **Moses**, a key figure in the Old Testament. Moses was born during a time when the Israelites were enslaved in Egypt. Fearing the growing population of the Israelites,

Pharaoh ordered the death of all Hebrew baby boys
(Exodus 1:22). To save him, Moses' mother placed him in a
basket and set him afloat on the Nile River.

Through God's providence, Pharaoh's daughter discovered
the infant and raised him as her own in the royal household
(Exodus 2:5–10). Moses grew up with the privileges of an
Egyptian prince but God shows Moses hints of his Hebrew
roots. As he matured, he became increasingly aware of the
suffering of his people.

One day, Moses witnessed an Egyptian beating a Hebrew
slave. Outraged, he intervened and killed the Egyptian
(Exodus 2:11–12). Fearing the consequences, Moses fled
Egypt and settled in the land of Midian, where he spent
forty years as a shepherd—far from Egypt and seemingly
far from the purpose God had planned for him.

But God had not forgotten Moses. One day, while tending
sheep near Mount Horeb, Moses encountered a burning
bush that was not consumed by the fire (Exodus 3:2).
There, God revealed Himself and called Moses to return to
Egypt to lead the Israelites out of slavery.

At first, Moses resisted. He felt unqualified, fearful, and
unsure of his ability to lead. He said to God, *"Who am I
that I should go to Pharaoh and bring the Israelites out of
Egypt?"* (Exodus 3:11). But God responded with a
powerful promise: "I will be with you" (Exodus 3:12).

With God's assurance and power, Moses returned to Egypt.
He confronted Pharaoh repeatedly, and through a series of
miraculous plagues—ten in total, each demonstrating
God's authority over Egypt's gods—Pharaoh was
eventually compelled to release the Israelites (Exodus 7–
12).

Moses then led the Israelites out of Egypt, through the parted Red Sea, and into the wilderness (Exodus 14). Over the next forty years, he served as their spiritual leader, lawgiver, and intermediary between the people and God. He received the Ten Commandments on Mount Sinai (Exodus 20) and guided the Israelites through seasons of rebellion, hardship, and divine instruction.

Moses was not perfect—he struggled with anger, doubt, and even disobeyed God at times (Numbers 20:7–12). But he remained faithful to the calling God placed on his life. His obedience and leadership shaped the identity of Israel as a nation and laid the spiritual foundation for generations to come.

Moses' story teaches us that discovering and fulfilling our God-given purpose isn't always straightforward. It may require leaving behind comfort, confronting past failures, and stepping into unfamiliar territory. But it also shows us that when we trust God and follow His leading, He equips us with everything we need—even when we feel inadequate.

So don't be afraid to dream big. Don't hesitate to ask God to reveal the unique purpose He has for your life. And don't be discouraged by setbacks or uncertainty. Remember, every step in your journey—including the detours—is part of God's plan.

You are not alone in this walk. God is with you—guiding, strengthening, and cheering you on. He has a purpose for you that exceeds what you can currently imagine, and as you walk in faith, you'll experience the joy and fulfillment that come from living in alignment with His will.

As we continue exploring the nurturing of the soul and the discovery of purpose, let us remember that purpose is not static—it evolves. What God calls you to today may grow and change over time. What matters is not having all the answers, but being open to God's leading.

In the next chapter, we'll dive deeper into the exploration of faith and spirituality—and how these two vital elements can enrich our understanding of the soul's journey.

Chapter 3
Exploring Faith & Spirituality

My Profound Experience at the Sea of Galilee

I would like to share with you the most profound moment of my life—one that occurred when I least expected it. On the morning of April 17, 2006, while continuing my visit to the Sea of Galilee, I asked the hotel clerk to arrange a taxi to take me to the church commemorating where Jesus fed 4,000 people, and then to Capernaum, the town that served as the headquarters of Jesus's ministry. The evening before, I had visited the Mount of Beatitudes, traditionally believed to be the site of the *Sermon on the Mount*—Jesus' most famous teaching and one of the most influential discourses in human history.

That morning, all the taxis I had seen were Mercedes, and since I had taken one back to the hotel the previous night, I had a general sense of the pricing. When I asked the hotel clerk to hail a taxi for a fair price, he offered to drive me himself. I agreed.

I expected a Mercedes, but his car turned out to be an old, beat-up vehicle cluttered with belongings in both the front and back seats. I could have refused or renegotiated the fare, but I sensed he could probably use the income, so I chose to continue the trip as planned.

After visiting the Church of the Loaves and Fishes in Tabgha, the driver mentioned a small church nearby on the way to Capernaum and asked if I'd like to stop there. I said yes.

The church was very small, peaceful, and empty—no other tourists were present. It was nestled quietly along the northwestern shore of the Sea of Galilee, right at the water's edge. Inside, the church was simple, with a central rock and birds flying freely within. After a few moments inside, I stepped back outside to take in the tranquil scene.

I asked the driver to take a photo of me as I walked out onto one of the partially submerged heart-shaped stones—about seven or eight feet into the calm, clear waters of the Sea of Galilee. After he snapped the photo, I bent down and gently touched the water with my fingertip.

The moment my skin made contact with the water, an overwhelming surge of energy entered my body. I was immediately filled with a deep and total sense of peace—something far beyond emotion or thought. I had never experienced anything like it before or since. It was as if the Holy Spirit had entered me, and I knew without a doubt that I was feeling the tangible presence and peace of God. The tingling in the finger that touched the water remained for nearly 30 minutes afterward.

That sense of peace—the purity, stillness, and connection to something divine—was the most profound spiritual experience of my life. Had I turned down the hotel clerk's unexpected offer, I would have missed that encounter entirely. It was a reminder that God's grace often meets us in the unlikeliest of ways.

Two weeks later, back in the United States, I began researching that little church. To my astonishment, I discovered it was the Church of the Primacy of Saint Peter, located in Tabgha. Built on the very shoreline of the Sea of Galilee, it marks the sacred site where Jesus appeared to His disciples for the third time after His resurrection (John

21). It was here that Jesus helped them catch 153 fish, prepared a meal of bread and fish, and restored Peter, saying, *"Feed my sheep."*

The rock I had stood on was one of twelve heart-shaped stones placed along the shoreline to commemorate the Twelve Apostles. I had unknowingly stood at one of the most sacred sites in Christian tradition—where the resurrected Christ reconnected with His closest followers and reaffirmed Peter's calling as the foundational "rock" of His Church.

The Church of the Primacy has welcomed notable pilgrims over the years, including Pope Paul VI in 1964 and Pope John Paul II in 2000. Yet, it hadn't appeared on my tourist map, and I had never even heard of it before. It felt like a divine appointment—one that God had orchestrated outside of my plans.

That day, I experienced firsthand the "peace of God, which surpasses all understanding" (Philippians 4:7). I felt like I was one with God—reconnected, realigned, and utterly still inside. The profundity of that moment—standing where Jesus appeared after conquering death—left a lifelong imprint on my soul.

Had I ignored the driver's suggestion or dismissed the church as unimportant, I would have missed the most spiritually transformative moment of my life. My soul was renewed by the living waters of Galilee, and my faith was reinvigorated in a way that no words can fully capture.

It was a sacred encounter that reminded me that God's grace often comes disguised in humble moments. That single experience not only uplifted my soul but also fortified me for the journey still ahead—with peace, with clarity, and with a deep awareness that I am not walking alone.

The Gospel Account of Jesus' Third Appearance

The Bible verses recounting Jesus's third appearance to his disciples after his resurrection at the Sea of Galilee can be seen in the Gospel of John 21:1-25. This account provides the context for the profound spiritual significance of the location where I had my unexpected encounter with God's peace.

Unifying Metaphor: Life as a Spiritual Walkabout

Just as the biblical patriarchs—Abraham, Moses, and even Jesus himself—were called into the wilderness to encounter the Divine firsthand, away from the constraints, noise, and distractions of society, my own life has unfolded as a kind of perpetual spiritual walkabout. With each new vista that has transfixed my senses—whether through travels across continents or the deep inward turns of self-reflection—another veil has lifted, revealing deeper clarity about the sacred architecture of existence.

From the hallowed shores of the Sea of Galilee, where the living waters first awakened my soul, to ancient desert petroglyphs etched by Indigenous societies millennia before the biblical scrolls were codified, this walkabout continues to affirm the same universal truths echoed across all enlightened traditions—that we are infinite souls journeying through space and time toward union with our Divine Source.

The temporary wanderings of this earthly exile—punctuated by oases of insight, rest, and divine encounter—constitute merely checkpoints along a much greater homeward sojourn. The challenge lies in remaining faithful to the course, even as mirages of material wealth, power, and fleeting pleasures offer detours so deceptively alluring that the disoriented soul can spend years—sometimes entire lifetimes—lost in the labyrinth of egotism and spiritual apathy.

It requires the eyes of unshakable faith, honed through adversity, to recognize the true path. It requires one to ignore distorted human maps of success and self-worth and instead be guided by the moral compass etched into the soul by the Celestial Navigator. Such has been the shape of my own personal exodus—a winding ascent marked by euphoric moments of divine intimacy, counterbalanced by painful descents into doubt and spiritual drought.

Exploring the Spiritual World

I've always been drawn to places of profound spiritual and physical elevation—from the Bible to the mountains. Among them, Mount Everest has long captivated my imagination and spirit.

To reach the start of the Everest Base Camp trail, I had to fly from Kathmandu, Nepal's capital, to Lukla—home of what is widely considered the most dangerous airport in the world. From the small aircraft's window, I could see the snow-capped summit of Everest itself, wind whipping snow off its peak into a majestic triangular plume stretching eastward, as though the mountain were breathing.

Lukla's Tenzing-Hillary Airport features a runway just 527 meters (1,729 feet) long and 20 meters (65 feet) wide, sloping steeply at a 12-degree gradient. Perched at 9,334 feet (2,845 meters) above sea level in the Himalayan cliffs, the margin for error is nearly nonexistent. As a private pilot myself, I had immense respect for the pilots who navigate such conditions. Too low, and the aircraft crashes into the mountainside; too high, and it overshoots the precipitous runway with concrete structures at the end of the strip.

From Lukla, the trek towards Everest Base Camp began—a journey that tested my endurance and stirred my spirit. The trail wound through narrow paths with cliffs on one side and swaying rope bridges suspended hundreds of feet above glacial rivers, with each step accompanied by panoramic Himalayan vistas that seemed almost surreal in their beauty.

While in Kathmandu, I also explored sacred Buddhist sites, including the iconic Boudhanath Stupa—a UNESCO World Heritage site and one of the largest and holiest stupas in the world. As I joined a few pilgrims circling its white dome clockwise in reverence, I could feel the centuries of prayer, devotion, and spiritual resonance embedded in its stones. I also sense the energy under my feet. Said to house relics of a past Buddha, this monument radiated an atmosphere of serenity and wisdom.

Whether standing beneath Himalayan peaks or walking the sacred mandalas of Nepal, I felt again that quiet calling— the soul's yearning for truth, connection, and meaning. Each location, each journey, offered more than scenic beauty—it became a spiritual checkpoint, a reminder of the shared longing within all humans to touch the eternal.

My Journey from Buddhism to Christianity

When I was young, I spent a week or two living as a novice Buddhist monk in a monastery alongside my father, uncles, and cousins. We shaved our heads, donned saffron robes, and committed ourselves to the principles of Buddhist monastic life, taking part in a temporary vow of renunciation. It was an educational and peaceful experience—one that helped me understand and appreciate the core teachings of Buddhism, particularly the Four Noble Truths, which explore the origins of suffering and the path to liberation and enlightenment.

My parents were devout Buddhists for over 80 years, deeply rooted in their faith and community. But in the summer of 1981, while I was on an engineering internship with U.S. Borax in the Mojave Desert, they surprised me. For the first time, they encouraged me to attend a Christian spiritual retreat being held by a Burmese Christian group at Lake Cachuma, near Santa Barbara, California.

That retreat turned out to be a turning point in my spiritual life. On the drive home to California City along the Pacific Coast Highway (Highway 101), my mind replayed many of the Bible verses I had heard over the weekend. When I reflected on Jesus' words to Peter in Matthew 26:34— *"Before the rooster crows, you will disown me three times"*—I was struck with overwhelming emotion and broke down in tears, though I was also filled with profound peace. At that moment, I surrendered my life to Christ and accepted Him as my Lord and Savior. Truly, God moves in mysterious ways, and His call often comes in the most unexpected of moments and places.

For decades after that initial moment of being born again on Highway 101, my sister Grace and I worked faithfully to share the Good News of Jesus Christ with our Buddhist parents. Grace regularly took them to Chinese-language church services and Bible studies. However, year after year, they remained polite but firmly committed to the spiritual path they had followed for most of their lives.

That changed in the final years of my father's life. A few years before his passing from COVID-19, my father—by then in his 80s—opened his heart. About two months before died, I vividly remember sitting with him when he said, through tears, that he felt Jesus was calling him home. It was one of only two times in my life I saw him cry—the first being at his brother Frank's funeral. He thanked me. I told him it wasn't necessary. I thanked him—for his love, his sacrifices, and the life he gave our family. I told him I loved him.

On Earth Day, April 22, 2020, six days after contracting COVID-19, my father passed away peacefully in a hospital a few miles from a five-star nursing home, where he lived due to mobility issues. I believe that day was no coincidence. Less than a year earlier, we had lost my brother—a gifted biotech and e-commerce entrepreneur— to heart disease at UC San Diego Hospital. I miss them both dearly.

Two years later, I saw a book on my mom's bookshelf titled 《标竿人生》 (The Purpose Driven Life) by Rick Warren—translated into Chinese. I asked her to read the title aloud, and she did with interest. I thought her dementia might prevent her from understanding much, but I was wrong. She had only forgotten some English words, not Chinese. When I asked my sister Grace how the book had arrived in our mother's home, she said she didn't know.

Then I remembered—I had bought one copy for each of my parents nearly two decades ago, quietly hoping it might plant a seed. The seeds we plant in faith may take years to bloom, but God's timing is always perfect.

To my amazement, my mom read the entire book, sometimes aloud, page by page, over the course of two months. I even took videos of her reading. She also read passages from the Bible, including John 3:16, which she particularly loved: *"For God so loved the world, that he gave his only Son, that whoever believes in him shall not perish but have eternal life."*

During a family celebration in Grace's backyard, I shared with my sister how Mom seemed increasingly drawn to the teachings of Christ. Grace gently asked our mother if she would like to accept Jesus as her personal Savior. Mom answered clearly and confidently: "Yes." Tears of joy flowed freely. Her spiritual rebirth was a moment of pure grace—a gift we had prayed for over decades.

As we continue exploring faith and spirituality, it's important to remember that every soul's journey is unique. Mine took me from the quiet halls of a Buddhist monastery to the transformative embrace of Christ, but the common threads—love, compassion, purpose, and hope—remain universal.

Chapter 4
Profound Losses and Reflection

When I first visited the desert town of Nazca in southern Peru—famous for the ancient and mysterious Nazca Lines, a UNESCO World Heritage Site etched across the arid plateau—I experienced an unexpected and deeply moving encounter. After going up to the hotel rooftop terrace for breakfast, I noticed a large banner celebrating the 100th anniversary of Rotary International. As a longtime Rotarian back home, I was stunned by the coincidence.

Intrigued, I walked into a nearby shop and asked how I could connect with the local Rotary Club. One inquiry led to another until I was introduced to some of the just 15 Rotarians who made up the entire Nazca chapter. I had never encountered such a small, grassroots club—a stark contrast to my home chapter in Saratoga, California, which had over 100 members.

To my surprise, they welcomed me and invited me to attend their weekly meeting. Afterwards, several members offered to take me on a tour of their current service projects. I accepted—and what I witnessed over the next several hours was truly transformative.

One site, in particular, left an indelible mark on my heart. We visited an impoverished community on the outskirts of Nazca, where a single elevated water tank provided the only clean water source. As we drove up the dusty hill, I noticed many small crosses embedded in the ground. When I asked about them, the Rotarians told me that many infants and children had died due to the lack of clean water and access to healthcare. Their families, unable to afford burial in city cemeteries, laid them to rest along the roadside.

They then introduced me to some of the families in the neighborhood. I saw how they fetched water from the tank and stored it in 55-gallon drums, some of which had visible mosquito larvae floating in them. The standing water posed serious health risks, including vector-borne diseases like dengue and malaria, which are prevalent in regions with poor sanitation infrastructure. The families explained how children often had to leave school early to begin working in fields or informal labor just to support their households.

Quietly, I learned that some of these children were involved in exploitative or dangerous work. In that moment, I was overcome by the most profound sense of grief I had ever felt for the injustice and suffering in the world. I couldn't hold back tears—I cried for nearly half an hour in front of those around me, something I had never done in public before. That moment altered me permanently. I resolved then and there to help the children of Nazca.

Despite its size, the Nazca Rotary Club celebrated Rotary International's centennial with pride. A parade marched through the town square, followed by a celebration attended by several hundred residents. I was deeply honored to be invited to carry the Rotary anniversary banner in the parade and assist in raising their flag in the plaza. That evening, they had a band perform, and I was asked to give a speech, which I gladly did. None of these plans were on my itinerary—it was a divine appointment I would never forget.

When I returned to the U.S., I reached out to Santa Clara University's Campus Ministry, explaining that I wanted to offer three fully funded scholarships for students to travel to Nazca to teach English. Campus Ministry connected me to the Ignatian Center for Jesuit Education, which sent out a campus-wide announcement. Many students applied, and together with the staff, I interviewed the applicants.

Meanwhile, I communicated with the Rotarians in Nazca via email—with translation support from bilingual volunteers on both sides—to plan logistics and housing. In the end, we selected three passionate students, who traveled to Nazca that summer. They taught English to over 250 local children across the city, giving them not just language skills but hope.

When the students returned, they shared their experiences with me over a celebration dinner at a Peruvian restaurant. Later, I was invited to speak at Santa Clara University's Board of Fellows quarterly meeting to present the Nazca Project.

Father Paul Locatelli, then-President of Santa Clara University, later invited me to serve on the inaugural advisory board of the Ignatian Center. I accepted and served for 11 years, continuing through the presidency of Father Michael E. Engh, S.J. It was a humbling and rewarding opportunity to help shape the university's approach to faith-based global service.

My awakening to Christ-centered service was only deepened by another journey—this time to Savannah, Georgia. There, I came face-to-face with the brutal legacy of the Atlantic slave trade, which operated through Savannah's port for nearly five decades. The city's cobblestone streets and historic structures bore witness to this painful chapter in history. I was deeply moved by the lingering generational trauma it represented.

Despite international laws aimed at preventing the expansion of slavery into new territories, Savannah,

Georgia, played a despicable role as a major port in the transatlantic slave trade. It received thousands of enslaved Africans trafficked across the Middle Passage under unimaginably horrific conditions. The scale of human loss, generational trauma, and systemic dehumanization represented by those haunting historical accounts shook me to my core.

Equally sobering was learning about and witnessing remnants of the Trail of Tears. Hundreds of thousands of Native American men, women, and children were forcibly displaced from their ancestral homelands during the 1830s by U.S. government policy. Thousands perished from starvation, exposure, and disease along the treacherous routes—a state-sanctioned act of ethnic cleansing that remains one of America's gravest moral failings.

All of these encounters with injustice—whether historical or contemporary, personal or observed—have profoundly shaped my view of the world and my role in it. They have strengthened my faith, but also compelled me to speak out and work toward building a more just, compassionate, and morally conscious society.

In reflecting on all this, I'm reminded of what the Bible says about those who have "seared their conscience" (1 Timothy 4:2)—individuals who persist in lies, speak in hypocrisy, and suppress truth for personal gain. History warns us what happens when moral clarity is sacrificed for power.

In April 2024, Israeli polls showed that approximately 71% of the public wanted Prime Minister Benjamin Netanyahu to resign. Nonetheless, on July 25, 2024, he addressed a joint session of the U.S. Congress for the fourth time—the most appearances by any foreign leader in American legislative history. This came amid global outcry over Israel's military actions in Gaza under his leadership.

By mid-2024, reports from humanitarian organizations, including UN agencies and human rights monitors, estimated over 45,000 people—primarily women and children—had been killed in Gaza. More than 90,000 were wounded, and 1.8 million civilians were displaced. Basic needs like food, water, and medical supplies were severely restricted. Much of the munitions and weaponry used were supplied by the US, making it complicit in the devastation.

Netanyahu received 49 standing ovations from members of Congress—beneath the glaring words engraved on the chamber wall: "IN GOD WE TRUST." What does this say about the moral compass of our elected leaders? Roughly 50 lawmakers boycotted the speech—a small but notable act of protest. Meanwhile, as of June 2024, Congressional approval ratings had plummeted to just 16%, reflecting the public's disillusionment with their actions and priorities.

I've previously met Israeli soldiers—both men and women—during my travels in Israel. They were decent, sincere individuals, committed to protecting their homeland. I also have Jewish friends and colleagues, both in the U.S. and Israel, who are thoughtful, accomplished, and compassionate. My critique is not of a people, but of leadership that forsakes justice.

On July 26, 2024, ABC News reported that British Prime Minister Keir Starmer's office stated the U.K. would not obstruct the International Criminal Court's (ICC) legal proceedings. On November 21, 2024, the ICC formally issued an arrest warrant for Benjamin Netanyahu.

Chapter 5
Grappling with Major Injustices in Life

The Global Financial Crisis of 2007–2009, also known as the Great Recession, stands as a stark reminder of economic turmoil's far-reaching consequences. Originating in the U.S. housing market collapse and the proliferation of subprime mortgage-backed securities, the crisis triggered a global chain reaction that led to historic financial institution failures and government bailouts. The United States bore much responsibility for this cataclysmic event, which devastated international markets, cost tens of millions of jobs worldwide, and wiped out trillions in global wealth.

Among the casualties was my law firm, Heller Ehrman. With a rich San Francisco history dating back to its 1890 founding and offices spanning the world's major financial centers, Heller had grown into a legal powerhouse boasting 730 attorneys at its peak. In a cruel twist of fate, on September 15, 2008—the very day Lehman Brothers, one of Wall Street's most venerable institutions, declared bankruptcy—Heller Ehrman was set to merge with Lehman-affiliated law firm Mayer Brown. The ill-fated union never materialized, and by year's end, Heller itself had succumbed, filing for Chapter 11 bankruptcy protection.

Heller Ehrman's implosion sent shockwaves through the legal community and had dire consequences for my own company, Iris Biotechnologies. We estimate Heller's misconduct and negligence resulted in damages exceeding $100 million. When Heller submitted bankruptcy filings, Iris Biotechnologies was listed twice among unsecured creditors—indicating the firm was fully aware of its obligations to us but failed to notify us directly.

In a shocking disregard for legal ethics, Bankruptcy Judge Dennis Montali turned a blind eye to the incontrovertible fact that under U.S. patent law, attorneys are prohibited from abandoning a client during prosecution without proper substitution of counsel or explicit consent from the USPTO. The court acknowledged a letter from the USPTO denying Heller Ehrman's request to withdraw representation from Iris Biotechnologies, yet ruled as if Heller's unlawful disengagement letter served as sufficient notice to seek alternative counsel—a decision without legal justification.

The depths of Heller's misconduct were further exposed when the court confirmed that the firm had received a letter from the USPTO indicating that Iris Biotechnologies' application for its "Artificial Intelligence System for Genetic Analysis" patent was on the verge of being granted. However, as a direct result of Heller's failure to respond, the USPTO declared the application abandoned.

Critically, Heller Ehrman concealed all three USPTO communications from Iris Biotechnologies. This egregious breach of fiduciary duty caused irreparable damage— denying Iris the opportunity to protect and commercialize a potentially revolutionary genomic diagnostic platform. At the time, the patent could have conferred a significant first-mover advantage in a nascent field that later exploded in valuation.

In another deceptive move, Heller never informed us that Iris Biotechnologies had been listed as a creditor nor disclosed any deadline for filing a claim. Upon discovering the firm's gross negligence and concealment, we acted swiftly, submitting claims within the statute of limitations. Still, in yet another blow, Judge Montali denied our $100 million claim, depriving us of our constitutional right to a jury trial and shielding the firm from further liability.

This quagmire traces back to September 16, 1999, when Heller Ehrman—with our full consent—shared Iris Biotechnologies' comprehensive business plan with Kleiner Perkins Caufield & Byers, a top Silicon Valley venture capital firm. The following year, in August 2000, Heller filed our patent application for the AI system for genetic analysis. Notably, our chief competitor, Genomic Health, was founded around this time, after we had already disclosed our business plan to both Kleiner Perkins and Sequoia Capital. These VCs later became early investors in Genomic Health, which went public in 2005 and was eventually acquired for $2.8 billion in 2019 by Exact Sciences.

Given the overlapping timeframes and shared investor exposure, questions remain about how Iris' intellectual property may have been compromised or indirectly leveraged.

On January 29, 2009, Law.com published a damning article titled "Heller Ehrman Estate Can't Buy Malpractice Coverage." It reported that Bankruptcy Judge Montali denied Heller's request to purchase three years of malpractice insurance, arguing the $10.2 million price tag outweighed the risk to creditors. This decision further compounded the injustice we experienced. In denying malpractice coverage, the judge effectively foreclosed the ability of clients—like Iris Biotechnologies—to seek restitution for demonstrable harm.

The bitter irony of Judge Montali denying Iris Biotechnologies the chance to present its $100 million claim was not lost on those familiar with the case. This was a story not only of legal negligence but also of systemic failure—where those sworn to protect justice instead entrenched injustice under the guise of expediency.

It was this very ruling that left Heller without the necessary insurance to satisfy the $100 million in damages it owed to Iris Biotechnologies as a result of its egregious and well-documented malpractice.

Determined to seek justice, we appealed the bankruptcy court's patently unjust ruling to both the District Court and the Ninth Circuit Court of Appeals. In a final indignity, Iris Biotechnologies was denied even the basic right to an oral hearing—our pleas for fair, impartial review of the facts fell on deaf ears.

This legal stonewalling compounded the pain of losing our opportunity to recover $100 million in damages stemming from Heller Ehrman's clear malpractice. Even more egregiously, Heller had listed Iris Biotechnologies twice as creditors in its own bankruptcy filings—a tacit admission of financial liability.

Their malpractice devastated our company's ability to survive. On a personal level, I lost approximately $20 million across my brokerage and private investment accounts. It is easy for others to say, "learn, forgive, and move on," but when justice is denied despite overwhelming evidence, healing becomes far more difficult.

As I watch images of innocent families in Gaza mourning the loss of their children, parents, and loved ones—killed without cause—I feel their pain on a spiritual level. When lives are shattered and no accountability is offered, how can true healing begin?

Too often, the perpetrators of injustice remain untouched by the suffering they cause. They go on with their lives, never forced to confront the magnitude of the damage they've inflicted. In the case of Gaza, this cycle of unaccounted injustice has created generational trauma that will reverberate long after the last bomb has fallen.

Now, this war has begun spreading—to Yemen, Lebanon, Iraq, and Iran. The U.S. and some of its allies are complicit, having supplied the very weapons used to carry out the mass destruction. In response, Russia has pledged to supply advanced arms to Israel's opponents, threatening to tip the conflict into a broader war—one that could spiral into World War III.

Injustice has consequences. For my investors, and me the price was financial devastation. For the people of Gaza and conflict zones across the world, the cost is irreparable human loss—loved ones who can never return. Who will be held accountable for these killings? What recourse is available when the world's most powerful nations enable violence instead of stopping it?

The survivors of the Holocaust and their descendants have endured unimaginable suffering. But as I witness the scale of devastation in Gaza, I am forced to ask: are some among them now enacting atrocities of their own? Have the persecuted, in some cases, become persecutors?

This crisis is devastating not just for Palestinians—but for Jewish people as well. Their collective memory, rooted in millennia of exile and tragedy, now risks being tainted by state-led actions that contradict the moral legacy of "Never Again." How much longer will this bloodshed continue, and how much deeper will it scar both peoples?

And what about lawyers and judges who perpetuate injustice—who protects society when the legal system itself is broken? When those tasked with upholding truth instead twist it to shield the powerful, where does a person turn?

As I reflect on my life's journey, I am filled with gratitude for the blessings I've received—from humble beginnings in Burma to scientific innovation in Silicon Valley. Through every season—whether in academia, biotech entrepreneurship, or spiritual discovery—God's grace and the love of family, friends, and mentors have been my compass. Grieving is not weakness—it is a holy, necessary process when the soul has endured great loss. It is my faith in God that sustains me.

I have many questions still. Some may never be answered. But I hope that by sharing my story—both the triumphs and the heartbreak—I can inspire others to persevere with passion, purpose, and compassion. Let us never lose sight of what truly matters: faith, family, science, and an unshakable desire to make this world more just and humane. As I prepare for the next chapter in my journey, I do so with a heart open to what lies ahead. Even when the night feels endless, the dawn is always near.

When I reflect on the grave miscarriage of justice inflicted by Judge Dennis Montali in the legal battle between Iris Biotechnologies and Heller Ehrman, I am reminded of Pontius Pilate—who, despite knowing Jesus' innocence, condemned Him under pressure. In both cases, truth was sacrificed on the altar of convenience, and justice was denied to the innocent.

Chapter 6
Lessons from History and Culture

Life on earth is temporary. Egyptian pharaoh Thutmose III, often called the "Napoleon of Egypt," was one of the most militarily accomplished rulers in history. He led at least 17 military campaigns during his reign in the 15th century BCE and greatly expanded Egypt's empire, recording his victories on the walls of the Temple of Karnak. He accumulated immense spoils and tributes from conquered lands.

Ramses II, known as Ramses the Great, ruled for 67 years during the 19th Dynasty and is widely regarded as Egypt's most powerful and celebrated pharaoh. He commissioned grand temples, including Abu Simbel, and statues that projected his divine status. Despite their earthly wealth and power, these pharaohs believed the afterlife mattered more than the present, and their elaborate tombs in the Valley of the Kings reflect this worldview.

We're going to look at some of the richest people in history to see if you know who they are.

As of September 30, 2024, the richest person in history is not Elon Musk ($244 billion), Jeff Bezos ($197 billion), or Mark Zuckerberg ($181 billion). It is Mansa Musa, the 14th-century emperor of the Mali Empire. Mansa Musa was born around 1280 CE and ruled from 1312 to 1337. During his famous pilgrimage to Mecca in 1324, he traveled with a caravan reportedly numbering 60,000 people, giving away so much gold in Cairo that it caused inflation lasting over a decade. Mansa Musa's wealth is estimated to be over $400 billion in today's dollars, though his fortune is difficult to quantify precisely due to the vast gold reserves of the Mali Empire.

As of December 2024, Elon Musk is the wealthiest living person, with an estimated net worth of US $486 billion, according to the Bloomberg Billionaires Index, and $464 billion, according to Forbes. His fortune comes largely from Tesla, SpaceX, and X (formerly Twitter). Here are some of the wealthiest historical figures before Elon Musk:

- **Genghis Khan (1206–1227)** – founder of the Mongol Empire, controlling the largest contiguous empire in history. While his wealth is hard to quantify, he controlled immense resources.
- **Zhao Xu (1048–1085)** – known as Emperor Shenzong of Song, ruler of China during the Song Dynasty, which accounted for nearly 30% of global GDP at its peak.
- **Akbar the Great (1542–1605)** – ruler of the **Mughal Empire** at its height in India, known for religious tolerance and administrative reforms.
- **Amenhotep III (c. 1388–1351 BCE)** – father of **Akhenaten**, he presided over Egypt's diplomatic and architectural golden age.
- **Augustus Caesar (63 BCE–14 AD)** – Rome's first emperor, who transformed it into a powerful imperial state, reportedly controlling wealth equivalent to 20% of the Roman world's GDP.
- **King Solomon (970–931 BCE)** – the biblical king of Israel, famed for his wisdom and massive wealth, including gold imports equating to hundreds of tons annually, though much of this is symbolic and difficult to historicize.
- **Mansa Musa (1280–1337)** – as mentioned, considered the richest human in recorded history.

Most Americans today have likely never heard of Zhao Xu, Akbar the Great, or Mansa Musa, despite their immense global significance. This highlights how Western-centric education can obscure global history.

Have you ever heard of Emperor Shenzong of Song, who ruled during a time when China's economic and technological output was unmatched globally?

Now consider figures remembered not for wealth but for moral or spiritual influence:

- Jesus Christ, born in Bethlehem, lived humbly, died crucified by the Romans, and is revered for His resurrection.
- Siddhartha Gautama (the Buddha) gave up royal luxury to seek enlightenment and taught liberation from suffering through the Eightfold Path.
- Muhammad, born in Mecca in 570 CE, worked as a merchant and became wealthy through his marriage to Khadijah, but is best remembered for receiving the Qur'an, the holy text of Islam.
- Andrew Carnegie, who said, "The man who dies rich dies disgraced," gave away over 90% of his wealth, funding more than 3,500 libraries worldwide. His adjusted net worth peaked at $372 billion in today's dollars.
- Nikola Tesla, despite revolutionizing modern electricity and inventing the AC system, radio, and induction motor, died in a New York hotel room in poverty. His inventions underpin much of the technology we rely on today.
- Socrates, the ancient Greek philosopher, left no writings himself but taught Plato, influencing Western philosophy's foundations.

- Leonardo da Vinci, the Renaissance polymath, painted the Mona Lisa and conceptualized early prototypes of helicopters, tanks, and solar machines.
- Michelangelo, who sculpted David and painted the Sistine Chapel ceiling, produced works considered masterpieces of human creativity. I saw his sketches—surprisingly more intricate than the finished paintings—and was reminded of my own childhood drawings.

When I was in Chaco Canyon, New Mexico, I saw the remarkable architecture and celestial knowledge of the Ancestral Puebloans, dating from 850 to 1250 CE. Their buildings align with solar and lunar cycles and remain marvels of engineering.

The Great Gallery, located in Horseshoe Canyon, Utah, is one of the most magnificent pictograph panels in North America. It spans nearly 200 feet and features anthropomorphic figures with intricate designs. Some archaeologists believe the art dates back 1,500 to 4,000 years, made by archaic hunter-gatherer societies, predating the Ancestral Puebloans.

The Barrier Canyon Style figures, often appearing as larger-than-life spiritual beings, evoke themes of shamanism and cosmic observation. The Museum of Modern Art and Denver Museum of Nature & Science have featured reproductions of this art, emphasizing its aesthetic and anthropological value.

To see the original Great Gallery, I drove 47 miles on a rugged dirt road, then hiked five hours in temperatures exceeding 100°F (38°C), descending 780 feet (237 meters) at the start and climbing back at the end—all alone. But the effort was worth it. The silence and grandeur of that sacred site moved me deeply.

What is God calling you to do while on this earth? People tend to believe what aligns with their upbringing, experiences, and worldview. When two groups believe that the other's belief system is nonsense, meaningful communication becomes nearly impossible.

In 2023, the Pew Research Center estimated the global religious distribution as follows:

- **Christianity**: 2.382 billion (31.0%)
- **Islam**: 1.907 billion (24.9%)
- **Secular/Nonreligious/Agnostic/Atheist**: 1.193 billion (15.6%)
- **Hinduism**: 1.161 billion (15.2%)
- **Buddhism**: 506 million (6.6%)
- **Chinese traditional religions**: 394 million (5.6%)
- **Ethnic religions** (not otherwise categorized): 300 million (3.0%)
- **Other religions**: around 800 million (8.0%)

Those unaffiliated with any religion now comprise over 1 in 7 people worldwide. As of 2024, with the world population at approximately 8.1 billion, Christianity remains the largest religious group.

The non-religious population stands at around 1.13 billion (16%), including atheists, agnostics, and others who do not identify with any faith. Islam follows closely with 1.6 billion adherents (23%).

Religious conflict has plagued humanity for centuries. One of the most hotly contested religious sites is the Temple Mount in Jerusalem, sacred to both Jews and Muslims. It is the location of the First and Second Jewish Temples, the latter destroyed in 70 AD by Roman forces. Muslims revere the same site as Haram al-Sharif, home to the Dome of the Rock and the Al-Aqsa Mosque, considered the third holiest site in Islam.

The tensions between Jews and Muslims over this sacred space are part of a conflict stretching back more than 1,300 years, of which the Israel-Gaza war is only the most recent eruption.

Islam was founded by the Prophet Muhammad in the 7th century, with the Hijra (migration to Medina) in 622 CE marking the beginning of the Islamic calendar. Sunnis believe the rightful successors to Muhammad were the first four caliphs, while Shias hold that Ali, Muhammad's cousin and son-in-law, and his descendants were the only legitimate successors.

In 1517, Martin Luther, a German monk and theologian, famously nailed his 95 Theses to the door of the church in Wittenberg, protesting the Catholic Church's sale of indulgences. This act ignited the Protestant Reformation, which emphasized direct engagement with scripture. Many Protestant denominations developed a more literal interpretation of the Bible, especially books like Revelation, shaping apocalyptic theology still prevalent today.

Due to light pollution, we've become disconnected from the cosmos. Before 1920, astronomers believed the Milky Way was the entire universe. That changed when Edwin Hubble confirmed that Andromeda was a separate galaxy in the 1920s. His findings expanded our understanding of the universe and our place in it.

I remember vividly when I first saw Andromeda through a 22-inch telescope. It was a coincidence. While driving from San Francisco to California City during my engineering internship at U.S. Borax in Boron, CA, I decided to take a seemingly shorter route through Los Padres National Forest via Highway 33.

The map suggested a shortcut, but the winding mountain roads slowed my speed to 15 mph in some stretches. After more than eight hours of driving in pitch darkness, I was exhausted. With no cell phones or GPS back then, I knew there'd be no rescue if my car failed.

At a remote 8,800-foot summit, I pulled into a parking lot to rest—only to be greeted by a crowd of angry people shouting for me to turn off my headlights. I had inadvertently stumbled upon a gathering of astronomers. After they calmed down, I learned that they were part of an amateur astronomy club observing the crystal-clear night sky.

Many of them had Celestron 8-inch or homemade telescopes. One kind observer invited me to look through his 22-inch scope focused on Andromeda (M31). I peered into the eyepiece and saw, with my own eyes, a spiral galaxy over 2.5 million light-years away. I forgot my fatigue, energized by the awe of witnessing another galaxy.

Most amateur astronomers use telescopes with 8-inch mirrors or smaller. Beginners often use models like the Celestron NexStar 5SE, a Schmidt-Cassegrain telescope with a 5-inch mirror. Coincidentally, I once met the CEO of Celestron on a small boat while traveling upriver in the Philippines—we were the only tourists on board.

Thanks to powerful tools like the Hubble Space Telescope and the James Webb Space Telescope (JWST), astronomers have discovered that the observable universe contains between 200 billion to 2 trillion galaxies. Most galaxies are 1,000 to 100,000 parsecs in diameter (3,000 to 300,000 light-years) and are separated by millions of parsecs (megaparsecs) in distance.

We live in a vast, almost unfathomable universe. In the face of such scale, we are humbled—and reminded of how small yet significant our lives are.

Most people today live under artificial lighting, which has significant consequences for human health, wildlife, and the natural environment. Major contributors to environmental pollution include vehicle emissions, industrial discharges, open burning by farmers and rangers, forest fires, and emissions from residential and industrial chimneys.

Light pollution, caused primarily by outdoor artificial light, disrupts ecosystems, affects human circadian rhythms, and obscures our view of the stars. It is one of the least discussed but most pervasive forms of pollution.

In 2016, scientists published the World Atlas of Night Sky Brightness, a computer-generated map created using thousands of satellite images. The Atlas reveals widespread light pollution across North America, Europe, the Middle East, and large parts of Asia. Only a few remote wilderness regions, such as Siberia, the Amazon Rainforest, parts of the Sahara Desert, and the deep interior of Australia, remain largely free from artificial nighttime illumination.

Singapore, Qatar, and Kuwait rank among the most light-polluted nations, where nearly 100% of the population lives under skies too bright to observe the Milky Way. In fact, over 80% of the world's population—and 99% of Americans and Europeans—can no longer see the natural night sky due to persistent skyglow.

Humans and animals have evolved to rely on the natural light–dark cycle, governed by the 24-hour circadian rhythm. Exposure to light at night disrupts this biological rhythm, interfering with sleep, hormone production, and immune system function. Melatonin, a hormone critical for regulating sleep, is produced in darkness and suppressed by light, especially blue light, which is common in LED lighting and electronic screens.

Chronic exposure to artificial light at night (ALAN) has been linked to sleep disorders, increased stress, anxiety, fatigue, and even elevated risks for obesity, depression, and certain types of cancer. The American Medical Association (AMA) has formally recognized these risks and supports policies to reduce light pollution and encourage further research into its long-term health effects.

Blue light, in particular, emitted from smartphones, tablets, TVs, and LED bulbs, has been shown to significantly lower melatonin levels. Reducing blue light exposure before bedtime has become a key recommendation by sleep specialists.

The night sky is not only a scientific frontier but also a source of spiritual and emotional awe. Living in the San Francisco Bay Area, I could only see a handful of stars at night. But in Yellowstone National Park and remote regions of New Mexico, the sky came alive with thousands of stars and a breathtaking view of the Milky Way. It's a profound difference—one that many city dwellers have never experienced.

Chapter 6
Lessons from History and Culture

Lick Observatory, nestled in the Santa Cruz Mountains, just an hour's drive from my home in Saratoga, CA. The observatory, established in 1888, is a renowned research facility operated by the University of California and has played a pivotal role in astronomical discoveries. Lick's "Evenings with the Stars" programs invite the public to engage with space science firsthand. During my time at UC Berkeley, I chose to study astronomy—not because it was required, but because I wanted to understand the universe.

As we explore beyond our atmosphere, another form of environmental degradation emerges: space pollution. The growing number of decommissioned satellites, defunct rocket stages, and orbital debris now clutter Earth's orbit. NASA and global space agencies track over 20,000 large pieces of space junk, while more than a million smaller fragments are believed to orbit the planet. Some interfere with telescopes, hampering our ability to study the cosmos.

The uncontrolled expansion of satellites—especially from large-scale projects like Starlink—poses both observational and navigational risks, and raises ethical questions about how we treat space as an extension of Earth's environment. Just as we've polluted our land, sea, and air, we now risk turning low-Earth orbit into a junkyard.

These concerns connect with broader themes—how the greatest empires in history rose and fell, how spiritual teachers have left legacies far more enduring than kings, and how nature's grandeur continues to offer insight and humility. Whether it's light pollution, space debris, or climate change, these problems point to a larger imbalance between modern human development and our spiritual, ecological, and ethical responsibilities.

Chapter 7
Personal Journeys and Reflections

Now, I want to share with you a two-week vacation in Alaska that was really refreshing for my soul. I saw many bears in Alaska, even a wild bear within 15 feet. It was pure joy to see a mother and calf Humpback whale breaching simultaneously several times from a ship sailing from Ketchikan, Alaska, through the Inside Passage to Anchorage. The ship stopped over at Juneau, where I visited the Mendenhall Glacier, a 13.6-mile-long glacier located in the Tongass National Forest. I took a helicopter ride to the top of Mendenhall Glacier, which is a mile wide. It was amazing to see virgin snow in the mountains. It was simply beautiful. I also walked to the edge of a crevice and looked down at the blue ice. I watched the water flow down several thousand feet.

It was pure joy watching Orcas (killer whales), the largest members of the dolphin family, which can reach speeds up to 34 mph, in the wild. They were simply majestic, unlike seeing their cousins in captivity performing for audiences. When I visited a fishery in Juneau, I saw many salmon swimming close to each other in the water. When I looked up, I thought, "Oh my God, this is the Pacific Ocean!"

In Ketchikan, I was walking around town and saw people fishing in a river from 30 feet above. What was unusual about this scene was that they were standing in the river and trying to catch salmon with fishing nets.

So, I asked some people what was going on. I was told that once a year, the authorities open the river for four hours for people to catch salmon with nets. This tradition is part of the community's subsistence fishing culture, and the temporary opening reflects state-regulated conservation efforts. It looked fun. So, I walked to a store that sells fishing gear and bought a net roughly 14 to 16 inches wide on a 6 or 7-foot pole, and started fishing. The river currents were strong, and the salmon moved fast. Catching a fish with a net was more challenging than I thought. The river was open for just four hours per year, and I felt lucky to be there.

After fishing for about fifteen minutes, a lady and her children came over and said, "You are having so much fun. You'll do better with a larger net. We are heading home now. You can use our net." I thanked them and took down their address. With the larger net, I fished for about half an hour. I caught a fifteen-pound salmon plus two smaller ones. I then walked to a gas station and asked how to get to the lady's house.

As it turned out, the map on the gas station wall did not include the newly developed street where she and her family lived. This reflects how some rapidly expanding communities in remote Alaska can outpace local infrastructure updates. The station attendant knew the general area and said, "You can take the map." I couldn't believe the lady and the station attendant's generosity. That would never have happened in California.

I decided to take the three salmon I caught to a fish market. They took out the roes and cleaned the fish. I then drove around to find the lady's address.

I returned her net and gave her the three salmon and roes. She was cleaning her fish when I got there. She then invited me to stay and enjoy salmon pasta with her family. I did, and we enjoyed our conversation. It was really refreshing to meet such kind people in Alaska.

Ketchikan averages about 153 inches of rain per year, making it one of the rainiest cities in the U.S. Once, I was talking to a young lady, and she said, "We are depressed because we haven't had any rain lately." I asked, "How long has it been?" She said, "3 days." I smiled. In California, droughts can last more than 20 years.

When I was in Haines, Alaska, where many eagles live, I took a picture of an eagle flying out of its nest. It looked just like the picture on a US postage stamp of an eagle with partially spread wings. The Bald Eagle population is particularly dense in Haines, where the Alaska Chilkat Bald Eagle Preserve is home to over 3,000 eagles each fall.

I also enjoyed Anchorage, Glacier Bay, Fairbanks and Denali Park. The 350-mile Denali Star Train, operated by Alaska Railroad, connects Anchorage to Fairbanks and offers stunning views from its glass-domed observation cars. I especially enjoyed Denali Park's Wonder Lake, from which you can see a very clear view of Mount McKinley, also known as Denali, the tallest mountain in the United States at 20,310 feet (6,190 meters). I had so much fun on my first Alaska vacation in 1994 that I forgot to eat for 24 hours.

On a different trip to Alaska with my parents, I took them to see a glacier that I saw on my previous trip from the visitor center. That was more than a decade earlier. Because of global warming, my parents and I could no longer see the glacier from the visitor center specifically built for viewing the glacier. We had to walk two miles to the glacier.

The path had signs showing where the glacier had receded annually due to global warming. This phenomenon is part of a well-documented pattern—Alaska has warmed more than twice as fast as the rest of the U.S. since the mid-20th century, according to NOAA.

While I was with my parents in Denali Park, we saw Bighorn sheep and many other animals. My parents and I also took a small plane to fly over Mount McKinley's base camp. We also flew very close to the summit, which was about 18 miles from the base camp, with a vertical gain of around 13,500 feet. Denali's base elevation is approximately 7,200 feet, and most summit expeditions start from the Kahiltna Glacier base camp. The views from the small plane were magnificent. My parents and I also saw many humpback whales and orcas in Alaska. I cherish the wonderful times I shared with my parents.

Just as the roots of a tree must strike deep beneath the surface to gather nutrients that feed the foliage, eventually reaching for the heavens, my own root system was anchored in academia and corporate facilities before ever branching out in service of more metaphysical practices to heal the soul. The material realms served as a fertile training ground for refining the discernment and skills that would later be redirected towards uplifting human consciousness itself through spiritual cultivation.

In this light, my work launching biotech with AI and medical companies dedicated to revolutionizing personalized medicine and analyzing genomic data can be seen as the transitional bridge—developing tools to map and optimize the body's biology was but a precursor for the realization that our minds and souls held the master codes still awaiting decryption. Having fulfilled my erstwhile pioneering roles in the physical domains, the final phase of reorienting fully towards the realm of spirit represented the ultimate culmination rather than a diversion from my path's overarching purpose.

For those whose spiritual senses have been awoken, we come to recognize our secular occupations and ambitions as merely support beams upholding the enlightened construction of our life's true magnum opus—an edifice of divine wisdom and transcendent service to be meticulously assembled unto glorification of the Creator's majesty. Only once our physical labors have erected a stable foundation can we then divert our full focus to the higher alchemy of spiritual work that truly elevates humanity towards its noble apex as bearers of God's light.

These personal journeys and reflections have shaped my understanding of the interplay between science, faith, and the human experience. They've taught me that our pursuits in the material world can serve as stepping stones to deeper spiritual insights and that every experience, whether in the laboratory or the wilderness, can be an opportunity for growth and connection with the divine.

Chapter 8
The Path Forward – Faith
and Perseverance

As we stand at the crossroads of history, technology, and spirituality, it's crucial to reflect on the lessons we've learned and chart a path forward that honors our spiritual heritage while addressing the challenges of our modern world. We are living in a time of unprecedented connectivity, scientific advancement, and social unrest, which makes the search for enduring meaning even more urgent. This chapter will explore how we can apply the insights gained from our journey to live more purposeful, faithful lives in the face of adversity.

The Importance of Faith in a Changing World

In a world that often seems to prioritize material success and technological advancement over spiritual growth, maintaining a strong faith can be challenging. Global surveys by institutions like Pew Research Center confirm a steady rise in secularism, especially in developed nations, yet spiritual longing persists across all cultures. Yet, as we've seen through the examples of historical and spiritual figures, it's often those who hold fast to their beliefs and values who leave the most lasting impact on the world.

In Matthew 16:26 (NKJV), Jesus said, *"For what profit is it to a man if he gains the whole world, and loses his own soul?"* This can be rephrased as: *"What will it profit a man if he gains the whole world and loses his own soul?"* This profound statement reminds us that our spiritual well-being should be our highest priority, even as we navigate the complexities of modern life.

In an age dominated by consumerism and instant gratification, this scripture serves as a timeless counterpoint to the fleeting rewards of worldly ambition. We must strive to balance our engagement with the world and our spiritual growth, always keeping in mind that our ultimate goal is not earthly success but alignment with God's purpose for our lives.

Perseverance in the Face of Injustice

As my personal experience with the legal system demonstrated, we often face injustices and setbacks that can shake our faith in human institutions. From wrongful convictions to systemic inequalities in bankruptcy and civil law, many people suffer outcomes that challenge their trust in man-made systems. However, it's in these moments of trial that our faith in a higher power becomes most crucial. Romans 8:28 says, *"And we know that for those who love God, all things work together for good according to His purpose."*

This doesn't mean that we should passively accept injustice. Rather, we should strive to be agents of positive change in the world, always guided by our faith and moral convictions. As Dr. Martin Luther King Jr. reminded us, *"The arc of the moral universe is long, but it bends toward justice."*

This quote, popularized during the Civil Rights Movement and originally paraphrased from a sermon by abolitionist Theodore Parker in the 19th century, continues to resonate in global struggles for fairness and reform. Our role is to actively participate in bending that arc, even when the path is difficult and the progress seems slow. Faith gives us the courage to endure injustice without becoming bitter and the strength to keep striving for righteousness when others lose hope.

Embracing Our Calling

Each of us has a unique calling, a purpose that God has designed us to fulfill. Discovering and embracing this calling is a crucial part of our spiritual journey. As we saw in the story of Moses, God often calls us to tasks that may seem beyond our capabilities. Exodus 3–4 recounts Moses' reluctance, citing his lack of eloquence and authority, but God empowered him through signs and the support of Aaron. With faith and perseverance, and by relying on God's strength rather than our own, we can accomplish great things.

Jeremiah 29:11 reminds us, *"For I know the plans I have for you," declares the Lord, "plans to prosper you and not to harm you, plans to give you hope and a future."* This promise can give us courage as we step out in faith to pursue our God-given purpose, even when the path ahead seems uncertain. Biblical history and modern testimony alike affirm that divine purpose often emerges most clearly in seasons of uncertainty and surrender.

Cultivating Compassion and Service

My experiences in Nazca, Peru, and elsewhere around the world have shown me the profound impact that compassion and service can have, both on those we serve and on our own spiritual growth. Jesus commands us to *"love your neighbor as yourself"* (Mark 12:31), and this love is most powerfully expressed through acts of service and compassion.

As we move forward, we must look for opportunities to serve others, especially those who are marginalized or in need. In Matthew 25:40, Jesus reminds us that whatever we do for the "least of these," we do for Him. This might involve volunteering, donating to worthy causes, or simply being more attentive to the needs of those around us in our daily lives. By doing so, we not only help others but also grow in our own faith and understanding of God's love. Compassion is not a passive emotion—it is love in action.

Stewardship of Creation

Our exploration of the night sky and the issues of light pollution remind us of our responsibility as stewards of God's creation. Genesis 1:28 gives humans dominion over the earth, but this is a call to responsible stewardship, not exploitation. The original Hebrew word for "dominion" (radah) in this context implies guardianship, not domination.

As we face environmental challenges like climate change and pollution, we must see our efforts to protect and preserve the natural world as an expression of our faith. Pope Francis' 2015 encyclical *Laudato* Si' echoed this, urging Christians to view care for the earth as a moral obligation. This might involve making more environmentally conscious choices in our daily lives, supporting conservation efforts, or advocating for policies that protect our planet. Protecting creation honors the Creator.

Nurturing Spiritual Growth in a Technological Age

As we've seen, technology has brought many benefits and challenges to our spiritual lives. The constant connectivity and information overload of our digital age can make it difficult to find quiet time for prayer, reflection, and spiritual growth. According to a 2023 Barna Group study, 56% of Christians reported that digital distraction frequently interferes with their spiritual practices.

Moving forward, we must be intentional about creating space in our lives for spiritual practices. This might involve setting aside specific times for prayer and Bible study, periodically "unplugging" from our devices to connect with God and nature, or using technology in ways that enhance rather than detract from our spiritual lives. Apps like YouVersion and Lectio 365 show how technology can be redeemed for spiritual formation.

Bridging Divides and Promoting Understanding

In a world often divided by religious, cultural, and ideological differences, we have a responsibility as people of faith to be bridge-builders. Jesus calls us to be peacemakers (Matthew 5:9), and this involves actively working to promote understanding and reconciliation across divides.

This might involve engaging in interfaith dialogue, working to address social injustices, or simply striving to show Christ's love to those who are different from us. Paul's teaching in Galatians 3:28—"There is neither Jew nor Greek... for you are all one in Christ Jesus"—points to the radical unity possible in the kingdom of God. By doing so, we can be a positive force for unity and peace in our communities and the world at large.

Preparing for the Future

As we look to the future, we must remain grounded in our faith while also being prepared for the challenges and opportunities that lie ahead. This includes staying informed about technological advancements and their potential impacts on society, engaging thoughtfully with ethics and policy issues, and continually seeking God's wisdom in navigating complex issues.

We should also be mindful of the potential for significant societal changes or challenges, as hinted at in biblical prophecies. Matthew 24, Revelation, and Daniel provide apocalyptic visions not to spark fear, but to encourage readiness, vigilance, and moral clarity. While we can't predict the future, we can prepare ourselves spiritually to face whatever may come with faith, courage, and hope.

As we conclude this journey through personal experiences, historical insights, and spiritual reflections, let us remember that our path forward is ultimately one of faith and perseverance. In a world that often seems chaotic and unpredictable, our anchor is our relationship with God and our commitment to living out His purposes for our lives.

The apostle Paul's words in Philippians 3:13–14 offer a fitting conclusion and challenge for us: *"Brothers and sisters, I do not consider myself yet to have taken hold of it. But one thing I do: Forgetting what is behind and straining toward what is ahead, I press on toward the goal to win the prize for which God has called me heavenward in Christ Jesus."*

Let us, too, press on, always seeking to grow in our faith, to serve others with compassion, to be responsible stewards of God's creation, and to be beacons of hope and love in a world that desperately needs it. May our lives reflect not just belief, but transformation—becoming living testimonies of the grace, truth, and power of God in a rapidly changing world.

Section 2: Take Actions While You Can

Chapter 9
The World at a Crossroads

Our world stands at a critical juncture, facing unprecedented challenges that threaten the very foundations of our global society. From the ongoing pandemic to the looming specter of climate change, from economic instability to rising social unrest, we are witnessing a convergence of crises that demand our immediate attention and action.

But these are not isolated developments they are deeply interconnected symptoms of a deeper fracture: a spiritual and structural unraveling of trust, truth, and equity on a global scale.

Perhaps most alarming is the growing chasm between the ultra-wealthy and the rest of society. As of April 2024, there were 2,781 billionaires globally with a combined worth equal to $14.2 trillion. The U.S. has 813 billionaires, followed by China, which has 473, and India, which has 200. This rapid accumulation of wealth at the top has far-reaching implications for social mobility, political influence, and economic stability.

The numbers are stark but more importantly, they reflect a global architecture where wealth accumulation has become disconnected from value creation, ethics, or human progress. Instead, the system rewards speculation, monopolization, and financial manipulation.

As of June 2024, the three richest individuals in the world Jeff Bezos, Elon Musk, and Bernard Arnault held a staggering $607 billion in wealth. This concentration of resources in the hands of so few stands in stark contrast to the struggles of billions who face economic hardship and uncertainty.

A handful of individuals now wield more economic power than many nations, influencing everything from markets and media to space exploration and climate strategies.

As of December 2024, Elon Musk is the wealthiest person in the world, with an estimated net worth of US$486 billion, according to the Bloomberg Billionaires Index, and $464 billion according to Forbes, primarily from his ownership stakes in Tesla, Inc. and SpaceX.

This era has produced techno-oligarchs entrepreneurs turned de facto sovereigns who often operate with little oversight and immense influence over public policy and collective imagination.

The power wielded by massive financial institutions further complicates this picture. BlackRock, the world's largest asset manager, oversees an astounding $11.5 trillion in assets as of November 2024. While positioning itself as a leader in environmental, social, and corporate governance (ESG), the company has faced criticism for investments in controversial industries and its close ties to the Federal Reserve during the COVID-19 pandemic. The sheer scale of BlackRock's influence raises questions about the concentration of economic power and its impact on global financial markets.

The very institutions that claim to champion sustainability and social responsibility are often deeply entangled in the engines of extraction, inequality, and political influence.

The three largest asset managers BlackRock, Vanguard, and State Street combined are the largest owners of 438 out of the 500 largest corporations in the US. These managers are also the largest shareholders in each other's companies. Since March 13, 2009, the S&P 500 has grown from 757 to 6,389 on July 25, 2025. Since March 9, 1990, the NASDAQ has grown from 437 to 21,108 on July 25, 2025.

Yet, this spectacular market growth primarily benefits the financial elite. The gains are concentrated, while the risks job losses, inflation, housing instability are distributed across the working and middle class.

Oligarchs and the Israel lobby control US politicians, and no matter who wins the US presidential election, Wall Street always wins ultimately. The wealthiest 10% of Americans own 93% of US equities, while the bottom 50% hold just 1%. From 1979 to 2019, the wages of the top 1% rose by 160% after inflation, while wages rose 345% for the highest 0.1% of earners. Wage growth in the middle has been sluggish, with median pay rising just 13.7%.

What we see is not capitalism, but a plutocratic shell where politics is theater and policy is engineered to maintain the wealth concentration. Democracy without economic justice is a hollow ideal.

Blackstone, with over $1 trillion in assets under management, has become the largest alternative investment firm globally and the largest commercial landlord in history. In the US, it is by far the largest landlord, owning almost 350,000 units of rental housing.

The roots of financial giants BlackRock and Blackstone can be traced back to the same institution that played a pivotal role in the 2008 global financial crisis, raising questions about the lessons learned from that near-catastrophic event.

Larry Fink, the CEO of BlackRock, and Ralph Schlosstein, two of BlackRock's founders, previously ran the mortgage-backed securities divisions at First Boston and Lehman Brothers, respectively. They initially joined Blackstone to manage an investment fund and provide advice to financial institutions.

This tight-knit circle of financial elites has not only recovered from past crises they've been rewarded for them. They now own more real estate, control more capital, and influence more government policy than ever before.

Blackstone was founded in 1985 as a mergers and acquisitions firm by Peter G. Peterson and Stephen A. Schwarzman, who had previously worked together at Lehman Brothers. Peterson was the former chairman and CEO of Lehman Brothers, and Schwarzman served as head of global mergers and acquisitions business. Schwarzman, the CEO of Blackstone, was briefly chairman of former president Donald Trump's Strategic and Policy Forum. Shortly after the 2020 election, President Biden hired two BlackRock executives for senior roles on his economics team. Brian Deese runs the National Economic Council, and Adewale Adeyemo is the deputy Treasury secretary. This revolving door between finance and government continues unchallenged, and bipartisan participation reveals that this is not a partisan issue it's a structural one.

The 2008 crisis, centered around the collapse of Lehman Brothers, brought the global financial system to the brink of total meltdown, requiring worldwide massive government bailouts, which some call the most significant generational theft in the history.

In 2009 and 2010, massive unemployment and sovereign debts threatened to destabilize economies, particularly in

Europe, resulting in civil unrest in various countries. Millennials, born from 1981 to 1996, were severely impacted by this generational theft. Their dreams of homeownership, job security, and upward mobility were shattered. Many became the first generation to expect a lower quality of life than their parents.

Wall Street CEOs and Big Bank CEOs were the most responsible for causing the Great Recession. President Barack Obama did not hold any of them accountable. None of them went to jail, including the CEO of Lehman Brothers, the fourth-largest investment bank in the United States that filed for Chapter 11 bankruptcy protection on September 15, 2008, with $613 billion in debt. Lehman was operational for 158 years, from its founding in 1850 until 2008. This impunity set a dangerous precedent: that systemic fraud, when done at scale, is immune from justice.

Lehman's bankruptcy filing remains the largest in US history and played a major role in the unfolding of the financial crisis of 2007–2008. Dick Fuld, the CEO of Lehman Brothers, walked away with half a billion dollars and three homes when Lehman failed. The US legal system did not hold him accountable for all the pain, suffering, and financial losses he caused. The collapse was largely due to Lehman's involvement in the subprime mortgage crisis and its exposure to less liquid assets. The ripple effects of this crisis continue to shape our economic landscape today. Many argue that the subsequent bailouts enabled by the Federal Reserve and policy decisions have only exacerbated inequality and set the stage for future instability.

In essence, the crisis wasn't resolved it was deferred, deepened, and reloaded into a new system of dependency.

The Federal Reserve has also enabled the financing of wars since WWI. US military spending rose from hundreds of millions pre-WWII to $85 billion in 1943 and $91 billion in 1944 with the Fed's help. The Treasury and the Federal Reserve devised plans for financing the war, meeting frequently to determine how to finance it through taxation and issuance of government bonds.

The 16th Amendment, ratified on February 3, 1913, established Congress's right to impose federal income taxes. Currently, the worst part of the unfair tax law is that it enables major corporations and very rich people to avoid paying taxes.

War and finance have long been partners funded by the labor of ordinary citizens, benefiting those with power, and framed as necessary for national security. But who really profits from war?

As we grapple with these challenges, it becomes increasingly clear that understanding our history is crucial to charting a path forward. The patterns of the past, the complex interplay of human nature and power dynamics, and the lessons of both triumphs and failures all offer vital insights for addressing our current predicaments. Learning about history is very important if we want to understand the present and create a better future.

History helps us avoid making the same mistakes again. In our fast-paced world, knowing history is more important than ever because many of today's challenges have deep roots in the past. But history is not merely academic it is personal. The choices of prior generations have built the systems we live under today, and we must choose whether to continue in their image or forge a new, more ethical path.

When we study history, we have to face hard truths so we can build a fairer world. History shows what drives social change. It also shows how resilient the human spirit is when facing impossible challenges. As we navigate the 21st century, history will be an essential guide. It gives us the context and perspective we need to understand our challenges and find effective solutions.

It reminds us of the power of people working together to make positive changes to our shared humanity. And it teaches us this: nothing changes without courage. No empire reformed itself from within without pressure. No elite ever surrendered privilege without resistance. Therefore, the people must rise not in rebellion, but in reformation.

Moreover, the psychological toll of these ongoing crises cannot be understated. The COVID-19 pandemic, in particular, has exposed and exacerbated existing mental health challenges, leading to widespread increases in anxiety, depression, and post-traumatic stress disorder.

The American Psychological Association's Stress in America surveys paint a picture of a nation grappling with prolonged stress and uncertainty. Collective traumas disrupt daily life, create feelings of uncertainty and fear, and challenge our sense of safety. The pandemic has increased rates of anxiety, depression, and PTSD across all age groups while making existing mental health inequalities worse.

Social isolation and disruption have added to this toll, showing how important social connection is for well-being. Mental health is not a secondary issue it is the foundation of a functional society. When the mind suffers, everything suffers: our creativity, our productivity, our empathy, and our moral clarity.

Economic hardship further compounds these mental health challenges. The shrinking middle class and growing income inequality create a pervasive sense of insecurity and despair. Families struggle to make ends meet, facing the constant threat of job loss or financial ruin. This stress ripples through relationships, impacting family dynamics and potentially setting the stage for long-term adverse childhood experiences. In the U.S., the share of adults in the middle class fell from 61% in 1971 to 50% in 2021.

At the same time, the share in the upper-income tier rose from 14% to 21%, and the share in the lower-income tier increased from 25% to 29%. In the past, some societies with these trends have collapsed. When prosperity is reserved for the few and despair becomes the default for many, social fabric unravels. This is how civilizations crumble not through bombs, but through broken homes and hopeless hearts.

The impact of economic hardship on relationships and family dynamics can be significant. Financial stress can strain even the strongest relationships, leading to conflict, resentment, and even divorce.

Children who grow up in families struggling with economic hardship may be more likely to face adverse childhood experiences, which can have lasting effects on their mental and physical health. This generational trauma is not theoretical it is measurable. It manifests in declining educational outcomes, poor health, increased crime, and reduced life expectancy. If we do not intervene, we risk raising a generation that inherits not only our debt but also our dysfunction.

As we stand at this crossroads, it's clear that business as usual is no longer an option. We must confront these challenges head-on, armed with knowledge of our past and a commitment to building a more equitable and sustainable future.

The path forward will require bold action, innovative thinking, and a renewed sense of global cooperation. To address the psychological impact of collective traumas and economic hardship, we must prioritize mental health in how we respond and recover.

This means making mental health services more accessible, providing support to vulnerable groups, and building communities that are more resilient. It also means dealing with the root causes of crises and working to create a society that values sustainability, fairness, and well-being. This is not merely a social agenda it is a spiritual one. Because to care for the soul of society, we must care for the souls within it.

In early April 2023, the National Opinion Research Center at the University of Chicago and the Wall Street Journal provided a side-by-side snapshot of the decline in the latest in a series of surveys tracking Americans' attitudes towards civic virtues, and the comparison is not flattering. "The NORC-WSJ survey reports, a lot has happened in America, but none of it is good."

In the last 25 years, the number of respondents who say that patriotism is "very important" to them has declined from 70 percent to 38 percent." For "religion," the number declined from 62 percent to 39 percent; "having a child" halved from 59 percent to 30 percent; "hard work" is down from 84 percent to 67 percent; and, after a brief spike in 2019, "community involvement" fell from 47 percent to 27 percent.

The only thing that has increased in importance for Americans, from 31 percent to 43 percent, is "money." We are witnessing a value inversion. What was once central to meaning faith, community, and purpose has been replaced by individualism, wealth accumulation, and digital distraction.

Most humans are slaves to money. It's okay to have money, but the love of money is the root of all evil. The Global Financial Crisis highlighted by the stock market crash of 2008 inflicted suffering on billions of people. The unbridled greed of Wall Street and The Great Recession forever negatively changed the global confidence in the US financial system.

Countries have started dumping the US dollar, which could ultimately lead to a new world order. Central banks have been buying gold in the past few years because they worry their fiat currencies will lose value. These movements signal not just market anxiety but also a shift in global trust and power. The world is hedging against American exceptionalism.

James Madison, America's fourth President (1809–1817), was known as the "Father of the Constitution." He said, "History records that the money changers have used every form of abuse, intrigue, deceit, and violent means possible to maintain their control over governments by controlling money and its issuance." This quote rings out across centuries as both warning and prophecy. We ignored it once can we afford to ignore it again?

Chapter 9
The World at a Crossroads

This is a wake-up call. People are trapped in complacency and addiction to the rising stock market, video games, drugs, television, the Internet, and other allures. Children are our hope, and we must rescue them. Because if we lose the minds and hearts of the next generation, we don't just lose culture we lose continuity. We lose the very idea of a better tomorrow.

The stakes could not be higher, but within these challenges are opportunities for transformative change. By understanding the interconnected nature of these issues from wealth inequality to mental health, from historical patterns to current crises we can begin to craft comprehensive solutions that address the root causes of our societal challenges.

The path ahead is difficult, but with collective effort, informed decision-making, and a commitment to equity and sustainability, we can work towards a future that offers prosperity and well-being for all. This is the moment. Not to panic but to prepare. Not to retreat but to rise.

Chapter 10
Federal Reserve and America's Economic Quandary

In the 1912 United States presidential election, Woodrow Wilson, a Democratic Governor from New Jersey with only two years of political experience, managed to defeat the incumbent Republican President William Howard Taft and the popular former Republican President Theodore Roosevelt, who served from 1901 to 1909. Wilson won with just 41.8% of the popular vote, the third-lowest winning margin in history, and he was the first Democrat to win a presidential election since 1892.

This victory was less about Wilson's popularity and more about a fractured political landscape. The Republican vote split between Taft and Roosevelt, handing Wilson the presidency despite a minority of support.

Taft and Roosevelt, who ran under the newly formed Progressive/"Bull Moose" party, split 50.5% of the popular votes. Roosevelt was shot on October 14, three weeks before the election, with a bullet that remained in his chest for the rest of his life. On October 30, 1912, Vice President James S. Sherman died, leaving Taft without a running mate less than a week before the election. Eugene V. Debs, the fourth-place finisher, won 6% of the popular vote, which remains the highest percentage ever won by a socialist candidate in U.S. presidential elections.

This chaotic election cycle unfolded during a time of deep political and economic unrest conditions that would soon reshape America's financial future.

Chapter 10
Federal Reserve and America's
Economic Quandary

In November 1910, six men, using only their first names, met secretly at the Jekyll Island Club to write the foundational plan for the Federal Reserve Act. For twenty years, they denied their meetings to draft the Aldrich plan, which formed the basis of the Federal Reserve, well after the key provisions of their plan were approved as the Federal Reserve Act.

The secrecy of this gathering, coupled with its immense impact on the future of American finance, has fueled debate and suspicion ever since.

Why did these six men, who were not members of the most expensive and exclusive club, have access to it for several days of meetings? Club members included prominent figures such as J.P. Morgan, Joseph Pulitzer, William K. Vanderbilt, Marshall Field, and William Rockefeller.

This begs a deeper question: how did a small group of mostly unelected elites come to wield such significant influence over America's monetary future, in a setting historically reserved for the financial aristocracy?

The six men included Republican senator Nelson Aldrich, chair of the Senate Finance Committee; Henry Davison, a partner at J.P. Morgan; Abraham Piatt Andrew, Assistant Treasury Secretary; Aldrich's private secretary Arthur Shelton; Frank A. Vanderlip, president of National City Bank, the largest bank in the U.S. and a former Treasury official; and Paul M. Warburg, regarded by many as the chief driving force behind the establishment of the Federal Reserve.

Together, these men represented a cross-section of government, finance, and industry entities now seen as entwined in shaping policy behind closed doors.

President Wilson appointed Warburg to the Federal Reserve Board from 1914 to 1918. Wilson opposed the Aldrich Plan because it gave the most authority to bankers. Wilson believed the plan needed oversight and that neither Congress nor the public would support a proposal that gave the government hardly any control.

Aldrich's bill faced strong opposition from politicians, who accused him of bias due to his close ties to wealthy bankers such as J. P. Morgan and John D. Rockefeller Jr., his son-in-law. Wilson's initial resistance reflected democratic concerns but even his version of reform would later be altered.

Although Congress rejected the "Aldrich plan," it laid the foundation for the Federal Reserve Act of 1913, which created the Federal Reserve System. The bill passed on December 22, 1913, and President Wilson signed it into law the next day.

The Federal Reserve Act that Wilson signed was altered after his signature, making it easier for the Federal Reserve to create money. This post-signature alteration weakened democratic checks and increased the Fed's autonomy, allowing it to operate more like a private entity than a government-controlled institution.

Wilson, lacking financial sophistication, likely didn't foresee how the Federal Reserve could evolve beyond his recognition. President Woodrow Wilson's deep Christian faith influenced him, and he ran for re-election in 1916 as the candidate to "Keep the US out of WWI."

Chapter 10
Federal Reserve and America's
Economic Quandary

Ironically, just a year later, America would be drawn into the very conflict he vowed to avoid triggered, in part, by the financial entanglements made possible through the system he helped create.

On October 15, 1915, American bankers organized under J.P. Morgan & Company authorized a $500 million loan to the British and French governments to finance their WWI expenses. On April 4, 1917, the U.S. Senate voted to support the measure to declare war on Germany, with the House concurring two days later. These were the initial steps for the U.S. to enter WWI. Financial interests, particularly through private lending to foreign governments, created a conflict of interest that made neutrality politically and economically unsustainable.

World War I resulted in approximately 40 million casualties, including 20 million deaths. I believe Wilson was a tortured soul burdened by guilt for his role in creating the Federal Reserve, which facilitated U.S. entry into WWI. He attempted to establish the League of Nations to prevent future wars but failed. In 1919, Wilson suffered a severe stroke that left him incapacitated until the end of his presidency in 1921.

Wilson's health decline seemed to mirror the crumbling peace he had tried too late to secure. His legacy remains deeply conflicted: architect of idealism, yet catalyst for mechanized warfare.

WWI and its aftermath led to WWII, in which an estimated 70–85 million people perished, accounting for about 3% of the global population in 1940. WWI and WWII also caused many people to become atheists. Prior to the creation of the Federal Reserve that facilitated wars, the United States stayed out of foreign war since the country's founding in 1776.

The 20th century introduced a pattern: centralized banking enabled militarized foreign policy, which in turn fractured spiritual and moral confidence worldwide.

If the Federal Reserve is allowed to facilitate WWIII, as it had done with WWI, WWII, and all the other wars that the US was involved in since WWI, billions of people will die. We must take action now to stop the Federal Reserve before it is too late. This is not hyperbole it is a warning grounded in historical precedent. Monetary policy and militarism have become twin engines of destruction, and history shows us the devastating cost of silence.

After the Stock Market Crash of 1929 and the economic problems that followed, millions of people suffered during the "Great Depression." According to Federal Reserve Chairman Ben Bernanke, the Federal Reserve's monetary policy played a significant role in causing the "Great Depression."

This acknowledgment from within the Fed itself reveals a haunting truth: that the institution's missteps can devastate entire generations.

The Federal Reserve also enables the rich to get richer and creates a wider and wider wealth gap between the very rich and the rest of the people. The Federal Reserve, a private bank and the central banking system of the USA, has enormous power and influence over the American and global economy. Yet, it operates with limited transparency and accountability.

Chapter 10
Federal Reserve and America's
Economic Quandary

The Fed's dual identity private in character, public in function makes it uniquely insulated from democratic scrutiny, even as it shapes the financial lives of every citizen.

Through its control of interest rates and ability to create money through quantitative easing (creating money out of nothing), the Fed's monetary policies have far-reaching impacts on economic growth, inflation, asset prices, and wealth distribution. Critics argue that the Fed's actions, especially after the 2008 crisis, have mainly benefited the wealthy by inflating asset bubbles while doing little for regular Americans. It rewards capital over labor, speculation over savings, and short-term profit over long-term stability.

The Fed has also played a critical role in financing U.S. government debt and deficit spending, effectively enabling the expansion of American militarism. The Fed helped finance World War I, World War II, the Korean War, the Vietnam War, the Middle East Wars, the Afghanistan War, the Ukraine War, the Israel-Gaza War, and other wars that took the lives of more than 100 million people and injured many more.

This unbroken chain of conflicts spanning over a century underscores how central banking has become a financial pillar of perpetual war.

Based on the Federal Reserve Economic Data, from 1901 to 1914, the average annual federal surplus was 5.79 million. After the creation of the Federal Reserve, from 1915 to 1963, the average annual federal deficit was 5.13 billion. Since the surge of the Vietnam War in 1964 to 2023, the average annual federal deficit was 409.38 billion.

During the Global Financial Crisis recovery years from 2009 to 2012, the average annual federal deficit was $1.27 trillion. During the COVID-19 recovery years from 2020 to 2023, the average annual federal deficit was $2.24 trillion. This exponential rise in deficit spending reveals a system addicted to debt that enables war, distorts markets, and burdens future generations.

As of July 25, 2025, our national debt was over 37 trillion. The interest payment alone on this debt exceeds $1 trillion annually, limiting the government's ability to invest in critical areas such as infrastructure, education, and healthcare. This debt crisis is not a distant threat it is a present reality already constraining our national priorities and eroding trust in public institutions.

The intertwining of the Fed with the interests of Wall Street and the military-industrial complex raises serious questions about whose interests it really serves. Critics argue that the Fed's low interest rate policies and quantitative easing after the 2008 crisis have contributed to asset bubbles, rising inequality, and increased risk-taking by financial institutions.

Lack of public understanding and democratic oversight of the Fed's operations is a major concern. Decisions that profoundly impact millions of lives are made by a small group of unelected officials, often with close ties to financial elites. True accountability requires sunlight. Yet the Federal Reserve remains cloaked in complexity, inaccessible to the average citizen, and virtually immune to meaningful reform.

The Federal Reserve controls the amount of money in circulation. Part of the money is grouped under M1 Money, which includes coins and currency in circulation +

checkable (demand) deposits + traveler's checks + saving deposits.

The other money group is called M2, which includes M1 + money market funds + certificates of deposit + other time deposits. These two categories form the backbone of monetary liquidity in the U.S. economy, directly affecting interest rates, credit availability, and inflationary pressures.

In 2023, the money supply in the U.S. shrank for the first time in 74 years. M2 money supply has dipped from a peak of $21.7 trillion in July 2022 to $20.87 trillion in December 2023. On Jan. 23, 2024, M2 is down a little over 2% on a year-over-year basis and 4.31% from its all-time high set in mid-2022.

When M2 shrinks, there are several implications. Interest rates rise. Economic growth slows. Unemployment increases. A shrinking money supply can trigger a chain reaction of economic pain particularly for small businesses, borrowers, and the working class while tightening access to capital across sectors.

Calls for Fed reform have grown louder in recent years. Ideas include increasing transparency and accountability, diversifying the Fed's leadership, and limiting its powers and role. Some argue for getting rid of the Fed altogether and returning to a money system backed by real assets. At the very least, we need a robust public debate about the proper role and governance of the Fed.

As we deal with the economic challenges of the 21st century, we must ensure our financial system serves the needs and interests of all Americans, not just a privileged few. A growing number of Americans now question whether a system designed over a century ago under vastly different economic and political realities can still meet the needs of modern society.

At the heart of America's economic challenges lies the Federal Reserve, a private institution with the extraordinary power to create unlimited currency for government borrowing and spending. This system, which places the burden of repayment on taxpayers, has undergone significant changes since its inception in 1913.

Understanding the role and evolution of the Federal Reserve is crucial to grasping the complexities of America's current economic situation. The Fed operates with unique and far-reaching authority blurring the line between fiscal and monetary policy in ways that directly affect national sovereignty and economic equity.

A pivotal moment came in 1971 when President Nixon severed the link between the US dollar and gold, transforming our currency into fiat money. This shift has had profound implications, not least of which is the skyrocketing national debt. The Federal Reserve's policies have enabled nearly perpetual wars and other costly endeavors, contributing significantly to this mounting debt.

The ability to create money without the constraint of gold backing has allowed for unprecedented levels of government spending, often with little regard for long-term fiscal sustainability. By removing the anchor of gold, the government removed its own restraints opening the door to debt-fueled policy decisions that benefit the few at the expense of future generations.

The erosion of the dollar's purchasing power is staggering. In 2024, a dollar is worth less than 4 cents compared to its value in 1913 when the Federal Reserve took control of the US banking system. To put this in perspective, prior to the Federal Reserve Act of 1913, an ounce of gold cost around $18.92.

As of April 22, 2025, that same ounce costs over $3,400 a stark illustration of the dollar's decline. This devaluation has significant implications for savings, investments, and the overall economic well-being of American citizens. Long-term savers are punished, wages lose real value, and financial security becomes more elusive particularly for the middle and working class.

This devaluation of currency is not without historical precedent. The hyperinflation in 1920s Germany serves as a cautionary tale. In early 1922, one US dollar was worth 160 German marks. By November 1923, this had plummeted to 4.2 trillion marks to the dollar.

While the US has not experienced hyperinflation of this magnitude, the 88% loss in the dollar's purchasing power since 1971 is cause for serious concern. It raises questions about the long-term stability of our monetary system and the potential for future economic crises. Unchecked money creation, whether driven by war, crisis, or politics, leads nations down a dangerous road. History has shown us where that road can end.

Had President John Kennedy not been assassinated in 1963, we may have avoided the high costs of the Vietnam War that later compelled Nixon to end the gold standard. The Vietnam War cost more than $120 billion and 58,220 American casualties.

Kennedy's economic vision focused on reforming the Federal Reserve and reducing military entanglements posed a direct threat to entrenched interests. His absence reshaped America's fiscal trajectory.

During the Great Depression, Congress enacted the Glass-Steagall Act to protect investors in 1933. The Banking Act of 1933 prohibited commercial banks from underwriting securities in order to prevent conflicts of interest and protect investors.

This law served the U.S. well until November 12, 1999, when the Gramm-Leach-Bliley Act repealed its provisions prohibiting bank holding companies from owning other financial companies. This repeal marked a turning point dismantling the firewall that had protected the economy from reckless speculation for over six decades.

The repeal of the Glass-Steagall Act dismantled this Depression-era law. The removal of this safeguard allowed for the creation of financial behemoths that were deemed "too big to fail," setting the stage for the moral hazard that would come to define the global financial crisis.

Also, while the Federal Reserve's monetary policy provided cheap money that fueled the real estate bubble, the government mandating lenders to issue more sub-prime mortgage loans accelerated the economic meltdown. This repeal significantly contributed to the near-collapse of the global financial system in 2008–2009. The cost of deregulation was paid not by those who engineered it but by millions who lost homes, jobs, and retirement savings.

If you research the causes of our key sufferings in the 21st century, it is quite evident that President Clinton is a key contributor to causing many of our miseries. First, he enabled the repeal of the Glass-Steagall Act, leading to the 2008–2009 Global Financial Crisis (Great Recession), resulting in the suffering of billions of people globally. His permissive handling of the Internet allowed the proliferation of pornography on the Internet. His push to expand NATO and NATO's ambition could eventually lead to WWIII in this decade. Clinton's economic and foreign policies widely praised at the time unleashed forces that have now become dangerously destabilizing.

In 1994, President Bill Clinton initiated the process to expand NATO. This process eventually resulted in the undeclared NATO-Russian war in Ukraine. Under President Biden, the United States Congress has approved more than $174 billion for Ukraine as of May 2024 to fight against Russia. Can you imagine the US government spending $174 billion to help homeless people within its borders instead? The comparison is sobering revealing where national priorities lie, and whose lives are deemed worthy of investment.

Citigroup and other firms spent $200 million lobbying for the repeal of the Glass-Steagall Act with the blessing of the President of the United States, Bill Clinton, the Chairman of the Federal Reserve System, Alan Greenspan, and the Secretary of the Treasury, Robert Rubin.

Between 1999 and 2009, Rubin received total compensation, including employee stock options, of $126 million from Citigroup. This revolving door between government and Wall Street is not just corruption it's legalized betrayal of the public trust.

In the wake of the Great Recession, the Federal Reserve and the US government's response bailing out big banks and insurance companies while forcing the public to absorb toxic debts has been a subject of intense debate.

The surviving banks emerged even larger and more powerful, raising concerns about the potential for an even more devastating crisis in the future. This concentration of financial power in fewer, larger institutions has implications for market competition, systemic risk, and the overall stability of the financial system. Instead of reforming the system, the crisis entrenched it further. Risk was socialized; reward remained privatized.

The scale of government spending and debt accumulation is staggering. The US military budget in 2023 stood at $847 billion, larger than the next ten countries combined. This accounted for about 13% of the federal annual budget of $6.4 trillion, which represented 23.83% of the US GDP. For comparison, in 2007, before the Great Recession, the US budget was $2.568 trillion, or 17.9% of GDP.

This dramatic increase in government spending raises questions about fiscal sustainability and the long-term economic health of the nation. We are now spending more than ever while growing weaker in core areas such as health, infrastructure, and education. This imbalance cannot last.

It's worth noting that in January 1835, for the first and only time in American history, all government interest-bearing debt was paid off. The contrast between this historical moment and our current situation is stark, highlighting the dramatic shift in fiscal policy and economic management over the past two centuries. From debt freedom to debt slavery this arc is both economic and moral.

Chapter 10
Federal Reserve and America's Economic Quandary

The challenges facing the US economy are complex and deeply rooted. Addressing them will require a clear-eyed assessment of the Federal Reserve's role, a willingness to learn from historical missteps, and the courage to implement sweeping reforms.

The Federal Reserve's dual mandate of maximizing employment and stabilizing prices has been criticized as conflicting goals that can lead to policy decisions that benefit some sectors of the economy at the expense of others. We must ask whether the Fed's objectives serve the people or merely preserve the status quo.

Moreover, the Fed's role in managing economic crises has expanded significantly since its inception. During the 2008 financial crisis and the COVID-19 pandemic, the Fed took unprecedented actions, including large-scale asset purchases (quantitative easing) and direct lending to businesses.

While these measures helped stabilize the economy in the short term, they have also raised concerns about moral hazard, asset bubbles, and the long-term consequences of such interventions. This trend of borrowing money to boost the US economy is unsustainable. Short-term fixes have become long-term habits and the bill is coming due.

The two largest US debt holders, China and Japan, have lost confidence in the US financial system. China's holding of US debt peaked in 2013 at $1.59 trillion, and Japan's holding of US debt peaked in 2014 at $1.53 trillion.

As of October 2023, Japan held United States Treasury securities totaling about 1.1 trillion U.S. dollars, and China held only $769.6 billion. In less than 10 years, Japan reduced its holding of US Treasury securities by $430 billion, and China sold over $768 billion. This retreat signals a global recalibration one in which the U.S. dollar's dominance is no longer guaranteed.

From Q3 2020 to Q4 2023, China's gold reserve went up by almost 2,000 tons, which is more than the UK, Japan, and India's gold reserves combined. It's like buying a quarter of the US gold reserves in three years. Only five countries have more gold than China. As of Q2, 2024, the US gold reserve was 8,133 tons, Germany's gold reserve was 3,352 tons, followed by Italy, France, Russia and China with 2,452, 2,437, 2,336 and 2,264 tons, respectively. Gold is making a comeback as a hedge against fiat collapse.

The story of the Federal Reserve and its impact on the US economy is inextricably linked to broader questions about the nature of money, the role of government in the economy, and the balance between short-term stability and long-term sustainability.

As we grapple with these issues, the need for financial literacy and civic engagement has never been more pressing. The decisions made in the coming years will shape the economic landscape for generations to come. The Fed is not just a financial institution it is a mirror of our national values, our priorities, and our collective future.

We must remember that the 2008 financial crisis brought the world to the brink of economic catastrophe; billions of people suffered, and the Millennials, especially, had to face diminished job prospects, stagnant wages, and mounting student debt. If we are to avoid another generational betrayal, we must act not with fear, but with foresight.

Chapter 11
America's Wars, Interventions, and Their Consequences

Since World War II, the United States has been involved in numerous wars and military actions across the globe. This pattern of intervention, enabled by the Federal Reserve's ability to finance deficits and fund military operations, has had far-reaching consequences both at home and abroad. Rather than being isolated incidents, these interventions form a continuous thread woven into America's post-war foreign policy a policy shaped by power projection, ideological containment, and resource control.

From the proxy battles of the Cold War to more recent conflicts in Afghanistan, Iraq, Syria, Ukraine, and Palestine, US military actions have often resulted in devastating human and economic costs for the countries involved.

While these interventions may have advanced some American strategic and economic interests, they have also raised questions about the true price of global hegemony. Each war leaves behind a trail of trauma, regional instability, and long-term resentment, often sowing the seeds of future conflict.

The ongoing "War on Terror," launched in the wake of the September 11, 2001 attacks, has seen the US become entangled in conflicts in Afghanistan, across the Middle East and Africa.

The 2003 invasion of Iraq, predicated on false claims of weapons of mass destruction, not only destabilized the region but also contributed to the rise of ISIS, creating new security challenges that persist to this day. These actions have left generations of civilians displaced or dead, infrastructures ruined, and extremism fueled not diminished.

After we invaded Iraq, the only thing we protected was oil. Iraq was the cradle of civilization, but museums with priceless artifacts were left unprotected for looters. One of the top profiteers from the Iraq War was the oil field services corporation, Halliburton. Halliburton gained $39.5 billion in "federal contracts related to the Iraq war."

Vice President Dick Cheney was previously CEO of Halliburton. This glaring conflict of interest epitomizes how modern warfare often enriches elites while devastating nations. It raises deep moral questions about whom war really serves and who pays the price.

In the ongoing Israel-Palestine conflict, the US has consistently provided billions in annual military aid to Israel, even as Israel's policies in the Occupied Territories have been widely condemned as violations of international law. This unwavering support has complicated efforts to resolve the conflict and has damaged America's credibility as an honest broker in the region. For decades, this policy has alienated much of the Global South, undermining America's reputation as a defender of democracy and human rights.

The recent escalation of violence following Hamas' attack on Israel on October 7, 2023, has brought these issues into sharp focus. The subsequent Israeli military response in Gaza has resulted in over 53,000 Palestinian death, mostly civilians, over 119,000 injured, and has displaced 1.8

million people as of May 7, 2025. Approximately 1,700 Israelis also died.

The conflict has spread beyond Gaza, involving Lebanon and Yemen, with the US and UK directly engaged in military actions against Yemeni targets. Israel has won almost every battle in Gaza, but it is losing the war in how it is being perceived by the vast majority of people in the world. This crisis has become a test not just of military strength but also of moral standing and the international tide is turning.

Critics argue that this pattern of militarism not only contradicts the principles upon which America was founded but has also eroded its moral standing in the world. The enormous financial cost of these interventions has diverted trillions of dollars away from pressing domestic needs while potentially contributing to the erosion of civil liberties at home. What America spends on projecting power abroad is often money not spent on education, healthcare, infrastructure, or its own veterans ironically, the very people sent to fight these wars.

The US has also been implicated in dozens of coups and regime changes around the world. For example, in 1967, Indonesian President Sukarno was forced to resign under pressure from a CIA-backed takeover led by Suharto.

Within a month, Suharto had signed away Indonesia's rights to the Grasberg mine, the world's largest gold mine, for free to a US company. This pattern reveals a familiar formula: a resource-rich country, a nationalist leader, and then a foreign-sponsored regime change favoring Western corporate interests.

However, in recent years, under President Joko Widodo's leadership, Indonesia successfully took over the Freeport Grasberg gold mine with 51% ownership, signaling a new era of assertiveness and sovereignty over its resources. This marks a rare case where a nation reclaimed what was taken under imperial leverage setting an example for others seeking economic self-determination.

Similarly, on September 11, 1973, the US supported Augusto Pinochet's military coup in Chile, which overthrew the democratically elected government of Salvador Allende. Pinochet's subsequent 17-year rule was marked by widespread human rights abuses, including the execution of thousands and the torture of tens of thousands.

At his death in 2006, about 300 criminal charges were still pending against him in Chile for numerous human rights violations. This intervention, justified in the name of anti-communism, left a lasting wound on Chilean society and set a grim precedent for Cold War-era foreign policy.

More recently, the US has become deeply involved in the conflict in Ukraine. In 2014, the US-backed coup in Ukraine shifted Ukraine's government from Russia to the West. This also resulted in Russia annexing Crimea and part of the Donbas region. This is similar to the US annexing Texas and California from Mexico in the nineteenth century. Geopolitical hypocrisy damages credibility what is condemned in others is often justified when done by the West.

Since the war began in 2022, NATO has supplied escalating lethal weapons to Ukraine and is currently risking a nuclear war. Each shipment of weapons prolongs the conflict and raises the stakes, pushing the world closer to a catastrophic confrontation.

NATO has also applied heavy sanctions against Russia, but they did not work as intended. More than a million soldiers and civilians have died in the war, and millions of Ukrainians are exiled.

Citizens in many EU countries are now paying much higher energy and food prices. The high price of fuel practically killed the German economy. The collateral damage of this war reaches far beyond the battlefield, crippling economies and shaking democratic stability across Europe.

Although Ukraine and Russia reached an agreement for peace in April 2022, the US, through the intervention of UK Prime Minister Boris Johnson, forced Ukraine to decline the deal. This is a tragic war that could have been easily prevented. The roots of many current conflicts can be traced back to past US-led NATO expansions eastward.

In 2025, US Secretary of State Mark Rubio and former UK Prime Minister Boris Johnson admitted on camera that the war in Ukraine was a proxy war against Russia. When leaders admit proxy motives after denying them for years, it confirms what many feared that this war is being fought not for peace, but for power.

The 1953 CIA-backed coup in Iran, which overthrew the democratically elected government of Mohammad Mosaddegh, set the stage for decades of tension between the two nations.

The subsequent support for the Shah's regime, followed by the 1979 Iranian Revolution and hostage crisis, has shaped US-Iran relations to this day. This episode illustrates how American interference often leads to long-term blowback, undermining the very stability it claims to protect.

The United States' conflict with Iran started when Britain's Prime Minister Winston Churchill asked for our help to overthrow the democratically elected government under Prime Minister Mohammad Mosaddegh in Iran. Britain was mad at Mosaddegh for his policy that nationalized the oil companies, and he was overthrown in 1953.

We helped Mohammad Reza Pahlavi, who had been King since 1941 and was widely known as the Shah, who was the last Shah of the Imperial State of Iran, to centralize power until his overthrow in the Iranian Revolution in 1979. In exchange for cheap oil and strategic partnership, the U.S. turned a blind eye to human rights abuses under the Shah until it could no longer ignore the consequences.

In return for providing the US with a steady supply of oil, the Shah received economic and military aid from eight American presidents. The Shah was superseded in 1979 by the theocratic government of Ayatollah Ruhollah Khomeini, a religious cleric who was exiled to France.

On November 4, 1979, Iranian militants stormed the United States Embassy in Tehran and took approximately seventy Americans as hostages. This terrorist act lasted 444 days. This pivotal moment reshaped U.S.-Middle East policy and continues to influence diplomatic tensions to this day.

The Cold War with the Soviet Union almost resulted in a global nuclear war during the 1962 Cuban Missile Crisis. The U.S. embargo against Cuba has been condemned by the UN for 32 consecutive years, with 185 countries demanding its lifting in 2023. Despite global opposition, the embargo persists, serving as a symbol of Cold War-era rigidity that no longer reflects current geopolitical realities.

This longstanding policy has been a source of tension in the Western Hemisphere and has complicated US relations with many Latin American countries. Instead of fostering cooperation, it reinforces distrust and anti-American sentiment in a region that craves mutual respect.

The 2003 U.S. invasion of Iraq, based on false claims of weapons of mass destruction, cost over $2.89 trillion and more than half a million lives. On April 24, 2024, President Biden signed a $95 billion war aid measure that includes aid for Ukraine, Israel, and Taiwan. These ongoing commitments to foreign military aid have raised questions about the prioritization of domestic needs versus international military engagements.

With soaring inflation, homelessness, and healthcare crises at home, many ask: Who benefits from this spending and at what cost to American society?

President Donald J. Trump did not start a new war during his first term. He has said prior to the November 5, 2024, US election that he would quickly end the war in Ukraine if elected president. His claim reflects a broader public fatigue with endless wars and a growing desire for a foreign policy centered on restraint and diplomacy.

As we reflect on this history of intervention and its consequences, it becomes clear that a fundamental reassessment of US foreign policy is needed. Approaches based more on diplomacy, conflict prevention, and respect for international law could better serve America's long-term interests and contribute to building a more peaceful world. The choices made in the coming years will shape the global landscape for generations to come. If the U.S. is to be a true leader in the 21st century, it must lead by example not by domination.

On the subject of conflict resolution, a long time ago, I was chosen to lead eleven newly formed teams comprising about 300 people at DuPont. One of the teams has worked together for many years and there was a great deal of animosity among approximately 25 people in the team. Though on a much smaller scale, their discord reflected the same human factors behind many global conflicts: pride, fear, and mistrust.

So, I asked them to name each of the grievances that they had and I wrote them down on big sheets of paper and put them up on the walls so that everyone could see. There were many pages, and the complaints totaled more than 300. Think about the conflicts you have had with someone or a group. How did you, if you did, resolve the conflict? Behind every war or dispute lies a human need recognition, dignity, clarity, or respect that went unmet.

As for the team above, they voted on each item and narrowed it down to the top 20 complaints, then 10, then 5, and ultimately 1. Some of the top issues involved honesty, trust, respect, communication, effort, balance of power, compromise, decision-making, retribution, limited resources, incompatible goals, unclear responsibilities, perception, pressure, outcome, and independence.

The number one cause of their conflict was respect. Once they started respecting each other more, the team felt better, performance improved greatly, and the bottom line improved significantly. The transformation was simple, yet profound: listen deeply, treat others with dignity, and focus on shared goals. These are not just corporate strategies they are blueprints for peace.

Chapter 11
America's Wars, Interventions, and Their Consequences

If you improve your "Problem-Solving Skills," you can go a long way. Whether in boardrooms or on battlefields, the true path to peace begins with the willingness to understand.

Chapter 12
Israel-Gaza Conflict – A Powder Keg
in the Middle East

The Israel-Gaza conflict, with its deep historical roots and complex geopolitical implications, stands as one of the most intractable and volatile situations in the modern world. The recent escalation of violence has brought this long-standing issue back to the forefront of global attention, raising urgent questions about peace, justice, and the future of the region.

It is not merely a regional crisis it is a global flashpoint that threatens to pull major powers into direct confrontation and destabilize the entire Middle East.

WWIII has not yet come, but danger looms in the air. In 1948, the Jewish State of Israel was established in Palestine, beginning a Jewish-Palestinian war that continues to the present day. In 1967, the Israelis captured Jerusalem in the "Six-Day War." Since then, repeated cycles of violence, occupation, and resistance have prevented any lasting peace.

The horrific attack by Hamas on October 7, 2023, which resulted in the deaths of 1,200 people in Israel, marked a new and tragic chapter in this conflict. Hamas, a Palestinian Sunni Islamist political and military organization that governs parts of the Gaza Strip, has long been at odds with Israel, but the scale and brutality of this attack shocked the world.

According to the U.S. State Department and the European Union, Hamas is designated as a terrorist organization. However, many in the Arab world and parts of the Global

South view it as a resistance movement against occupation highlighting the deep division in global narratives.

The Israeli response to this attack has been devastating. As of May 7, 2025, the Israel-Gaza war has claimed over 53,000 Palestinian lives, the majority of whom were civilians, including a large number of children and women. More than 120,000 people have been injured, and an astounding 1.8 million have been displaced. The humanitarian crisis in Gaza has reached catastrophic proportions, with widespread destruction of homes, infrastructure, and essential services.

Reports from international humanitarian organizations such as UNRWA and Médecins Sans Frontières confirm the collapse of healthcare, sanitation, and food access in Gaza, describing it as one of the worst humanitarian disasters of the 21st century.

The conflict has sparked global outrage and led to widespread protests, particularly on college campuses around the world. Demonstrators are demanding divestment from Israel and certain Israeli companies, leading to tensions with university administrations. Some institutions have begun expelling students, further escalating the situation.

In April 2024 alone, there were approximately 2,000 arrests related to these protests. Many student-led movements have invoked parallels to the anti-apartheid movement in South Africa, increasing pressure on institutions to take moral stances amid political controversy.

The international community has struggled to respond effectively to the crisis. On May 20, 2024, following a thorough investigation, the International Court of Justice announced its intention to seek the arrest of Hamas leaders for war crimes committed in Israel, as well as Israeli Prime Minister Benjamin Netanyahu for alleged war crimes in Gaza.

This development underscores the complex and contentious nature of the conflict, with atrocities committed on both sides. This is the first time in modern history that a sitting Israeli prime minister has faced such charges from an international body signaling a possible shift in global accountability standards.

The US role in this conflict has been a subject of intense debate. As a long-standing ally of Israel, the US has consistently provided massive military aid and diplomatic backing. This support has complicated efforts to reach a peaceful resolution.

The US has supplied Israel with over $150 billion in aid since its founding, including $3.8 billion annually under a 10-year memorandum signed during the Obama administration. Critics argue that unconditional support limits America's ability to act as a neutral mediator.

However, there have been moments of progress in the past. The 1978 Camp David Accords, mediated by President Jimmy Carter, led to a peace deal between Egypt and Israel. The 1993 Oslo Accords established the Palestinian Authority and seemed to offer a path towards a two-state solution.

Yet, the assassinations of Egyptian President Anwar Sadat and Israeli Prime Minister Yitzhak Rabin by hardliners opposed to highlight the challenges facing those who seek peace. Both leaders paid the ultimate price for choosing diplomacy over war underscoring how extremism on both sides has repeatedly derailed hope.

The US, given its influence and resources, has the potential to play a crucial role in shaping the future of this conflict. Advocates argue that it must use its leverage, including the billions in annual aid to Israel, to discourage actions that undermine peace, such as the expansion of settlements in the occupied territories. They also call for holding all parties accountable for violence and human rights violations. Many now believe a new framework is needed one rooted not in historical favoritism, but in international law and genuine equality.

The Middle East is increasingly interconnected by overlapping alliances and proxy dynamics, meaning one spark in Gaza can ignite flames across the entire region.

Since November 2023, the Houthis have attacked many commercial boats and U.S. ships. In most cases, the U.S. is launching $2 million defense missiles to stop $2,000 Houthis drones. The Houthis have also taken down 7 $32 million US Reaper drones in addition to diverting US warships, including a carrier, from Asia to Yemen. Moreover, US has lost 3 F/A-18 fighter jets, each costing about $67 million, for various reasons. This asymmetric warfare has exposed the vulnerabilities of traditional military might and further burdened U.S. taxpayers in a war with no clear endgame.

The conflict with Yemen in the Red Sea has increased freight costs, shipping time, and the expenses of insuring commercial trade goods. Lebanon has fired a ballistic missile into Israel for the first time after Israel's massive bombing in Lebanon. This war is escalating fast. The Red

Sea disruptions have already impacted global supply chains, pushing inflation higher and affecting food and fuel prices worldwide.

The conflict has also exposed deep divisions within the United States. While the powerful Israel lobby, including organizations like AIPAC, has maintained strong support for Israel in Congress, there has been growing criticism of Israeli actions, particularly among younger Americans and progressive politicians.

Recent polls by Pew Research show a significant generational shift: while 61% of older Americans sympathize with Israel, only 32% of Millennials and Gen Z do indicating a potential future realignment in U.S. foreign policy.

As the death toll mounts and the humanitarian crisis deepens, the need for a just and lasting solution becomes ever more urgent. Yet the path to peace remains elusive. Any resolution must grapple with fundamental issues of land rights, security, and self-determination for both Israelis and Palestinians.

A two-state solution, while long envisioned, now seems more distant than ever, as settlements expand and extremist voices grow louder on both sides.

The Israel-Gaza conflict serves as a stark reminder of the consequences of unresolved historical grievances and the dangers of entrenched positions. It highlights the need for courageous leadership, innovative diplomacy, and a willingness to challenge long-held assumptions.

As the world watches events unfold in this troubled region, the decisions made in the coming months and years will have profound implications not just for Israelis and Palestinians but also for global peace and stability. With the Middle East at a boiling point, inaction is no longer an option especially for a world that claims to value justice, human dignity, and peace.

As we look back on our own history, we can see that the genocide and ongoing ethnic cleansing in Gaza is a smaller, contemporary version of what the United States military did to the Native Americans. Though the contexts differ, the moral parallels force us to reflect: how can a nation born from overcoming injustice remain silent when others endure it?

However, the US has come a long way. The United States has the most generous and innovative people enjoying freedom that most people on Earth can only dream of. We must do our part to preserve that privilege in a fair way. The time to act is now. If America is to remain a beacon of hope, it must lead not by force, but by example. True leadership is not domination it is compassion, courage, and the pursuit of justice.

Chapter 13
NATO's Expansion and the
War in Ukraine

The ongoing conflict in Ukraine represents one of the most significant geopolitical crises of the 21st century, with far-reaching implications for global security and the international order. At its core, this crisis is deeply intertwined with the history of NATO expansion and Russia's perception of threats to its security.

The roots of the current conflict can be traced back to the end of the Cold War. On February 9, 1990, US Secretary of State James Baker made a crucial promise to Soviet leader Mikhail Gorbachev: NATO would not expand eastward if Russia accepted Germany's unification. German Chancellor Helmut Kohl and NATO's secretary general reiterated this assurance in the following months.

For Gorbachev, these assurances were pivotal in clearing the way for a compromise on German reunification. However, this promise was not formalized in the treaty signed on September 12, 1990. Declassified documents later revealed that multiple Western officials gave verbal assurances to Gorbachev, but no legally binding commitment was made.

Following the collapse of the Soviet Union in 1991, many former Warsaw Pact and post-Soviet states sought to join NATO. Despite Russian opposition, Poland, Hungary, and the Czech Republic became NATO members in 1999. This was followed by the accession of seven additional countries in 2004: Bulgaria, Estonia, Latvia, Lithuania, Romania, Slovakia, and Slovenia. Subsequent expansions included

Albania and Croatia (2009), Montenegro (2017), North Macedonia (2020), and Finland (2023).

Russian President Vladimir Putin, in his now-famous 2007 Munich speech, expressed Russia's concerns about NATO expansion: "I think it is obvious that NATO expansion does not have any relation with the modernization of the Alliance itself or with ensuring security in Europe. On the contrary, it represents a serious provocation that reduces the level of mutual trust. And we have the right to ask: against whom is this expansion intended?"

The situation reached a critical point in 2008 when NATO welcomed Ukraine and Georgia's aspirations for membership. Russia saw this move as crossing a red line. At the Bucharest Summit, NATO officially stated that Ukraine and Georgia "will become members of NATO," though no specific timeline was offered.

In 2014, a Western-backed coup in Ukraine led to a pro-Western government, further alarming Russia. The Ukraine war started in 2014 when Victoria Nuland, the assistant secretary of state for European and Eurasian Affairs, and neocons instigated the coup that illegally overthrew the Yanukovich government. Russian intelligence intercepted and leaked to the international media a Nuland telephone call with U.S. ambassador to Ukraine Geoffey Pyatt discussing in detail their preferences for specific personnel in a post Yanukovych government while Viktor Yanukovych was still President.

However, many scholars and Western analysts describe the 2014 change in government as the result of a popular uprising (Euromaidan) against widespread corruption and the rejection of an EU trade deal not an orchestrated coup.

The tensions finally erupted into full-scale war in 2022 when Russia invaded Ukraine, viewing it as an intolerable step towards NATO membership. The invasion was widely condemned internationally and met with Western sanctions and military aid to Ukraine.

Professor John Mearsheimer of the University of Chicago has consistently warned about the dangers of NATO expansion, arguing that it was the root cause of the war in Ukraine. He was a United States Military Academy West Point graduate, and he served in the US Army and Air Force for ten years. His analysis suggests that Russia would go to great lengths to prevent Ukraine from joining NATO, even if it meant decimating the country. In his 2014 article "Why the Ukraine Crisis Is the West's Fault," Mearsheimer argued that Western policymakers mistakenly assumed Russia would passively accept NATO's eastward expansion.

The conflict has drawn in major global powers. As of May 2024, the United States has been a key supporter of Ukraine, with Congress approving over $174.8 billion in aid since 2022. Of that amount, approximately $83.4 billion had been disbursed by early 2025, including military, humanitarian, and financial support (USA Facts). This involvement goes beyond mere support for Ukrainian sovereignty. According to US Senator Lindsey Graham, Ukraine possesses an estimated 10 to 12 trillion dollars' worth of critical minerals that could be used to repay the US.

While this figure is widely cited, independent estimates from the Ukrainian government and mining analysts suggest Ukraine's recoverable critical mineral reserves are valued in the lower trillions range significant, but difficult to precisely quantify.

Moreover, Ukraine has already sold off $400 billion in assets on Wall Street despite its national GDP being only $200 billion. This claim is difficult to verify and likely conflates long-term investment opportunities, future valuations, and speculative transactions rather than actual asset liquidation of that scale. Ukraine's privatization and foreign investment policies have opened its market to international entities, but exact figures remain contested.

The strategic importance of Ukraine extends beyond its mineral wealth. The US needs the critical minerals that Ukraine has for its military and industrial applications. Similarly, Taiwan, another flashpoint in international relations, is crucial for the US due to its advanced semiconductor industry, which is vital for military, artificial intelligence, and other high-tech applications. I visited most of the semiconductor companies, such as TSMC in Taiwan, for business during the late 90's.

There's a chance the Ukraine war could trigger WWIII, with the US and its allies on one side and Russia and its allies on the other. The West, led by the US and its allies, has given Ukraine hundreds of billions of dollars in cash, weapons, and other forms of assistance while its own citizens have suffered high inflation in food, energy, and other costs. Inflation rates in the US and Europe surged post-2022, partly due to the war's effects on global energy and grain markets. However, inflation has gradually declined in 2023–2024 according to central banks.

There have been hundreds of thousands of deaths in Russia and Ukraine, and millions have fled because of the Ukraine war. As of early 2024, estimates suggest over 500,000 soldiers have been killed or wounded on both sides, and more than 6 million Ukrainians have become refugees

(UNHCR). As a result of the disruption of oil and gas flow from Russia, many European countries are burning more coal, causing climate change to worsen. Germany, for example, reactivated coal plants to stabilize its energy supply, despite long-term commitments to renewable energy.

The US-led, NATO-backed coup in 2014 replaced the Ukrainian government loyal to Russia with one on the NATO side. Right after the coup, Russia annexed Crimea to protect its naval base there. Russia shares a 1,200-mile border with Ukraine, and it wanted to prevent NATO from installing weapons at this border. Russia also wanted to protect ethnic Russians in Ukraine, who were persecuted by the new regime.

So, Russia supported Russians in Ukraine's eastern region to claim independence and then become a part of Russia. International human rights organizations have noted tensions between ethnic Russians and the post-2014 Ukrainian government, though claims of systematic persecution have been heavily disputed and used by Russia to justify military intervention.

It didn't make sense as to why NATO and the EU seemingly sacrificed the welfare of their citizens by so easily giving up cheap gas from Russia and providing hundreds of billions of aid to Ukraine until you realize that NATO, EU, and the US may be interested in securing mining contracts from Ukraine.

Ukraine is home to a vast array of critical minerals, including the largest titanium reserves in Europe, accounting for 7 percent of the world's reserves, and almost 500,000 tons of lithium, with an estimated value in excess of US$26 trillion. Ukraine also holds significant untapped rare earth deposits, graphite, cobalt, and other minerals essential for green technologies and defense systems.

Ukraine's reconstruction is projected to cost $486 billion. In October 2024, Ukrainian President Volodymyr Zelenskyy announced a partnership with global asset manager BlackRock to coordinate international investment in rebuilding Ukraine's economy (Reuters). BlackRock's Financial Markets Advisory division is tasked with overseeing investor strategy, signaling that Western corporate and financial interests will likely secure the largest reconstruction contracts.

Germany's new Chancellor Fredrick Merz was formerly the Chairman of BlackRock Germany. He was also a senior counsel for Mayer Brown, the law firm for Lehman Brothers. Lehman was the most responsible for causing the near total meltdown of the global financial system in 2008. On May 14, 2025, Merz pledged that Germany would have the "strongest conventional army in Europe."

Zelenskyy's presidential term expired on May 20, 2024, yet he remains in power under wartime emergency law. This has raised constitutional concerns, especially since no election has been scheduled. Ukraine has at least five million refugees abroad, half of whom are eligible to vote, and an additional six million internally displaced persons. The logistics of organizing a national election under martial law and active war conditions remain unresolved. It is difficult to imagine how the hundreds of thousands of Ukrainian men and women currently serving in the armed forces could meaningfully participate in the election.

The Ukraine conflict has had global consequences. It disrupted energy markets, drove up inflation, worsened food insecurity in developing nations, and strained diplomatic relations between East and West. A clear contrast was seen in April 2024, when the US and its allies intercepted over 180 ballistic missiles and drones launched by Iran toward Israel, a demonstration of high-tech air defense and unified action. By contrast, Ukraine has endured daily missile and drone attacks for over two years with less coordinated Western military intervention to shield civilians. This disparity has led some to argue that Ukrainians are being treated as expendable geopolitical pawns rather than valued allies.

Critics assert that the US-led NATO alliance is more interested in Ukraine's vast mineral wealth than in the lives of its citizens. Ukraine possesses massive deposits of critical minerals, including the largest known titanium reserves in Europe, substantial lithium, cobalt, rare earth elements, and other materials crucial to defense, energy, and tech industries (BBC). The sooner the Ukrainian people understand this, the argument goes, the sooner they can negotiate peace with Russia and secure a future based on neutrality and non-alignment.

Resolving this conflict and the broader NATO-Russia divide requires a frank assessment of how actions on all sides escalated tensions. NATO expansion, Western support for Ukraine's political transformation, and Russia's military aggression have all contributed to this breaking point. The West's idealistic narrative about "freedom and democracy" masks the realpolitik motivations underlying the conflict, particularly resource acquisition and strategic positioning.

Professor Jeffrey Sachs of Columbia University, a Jewish American economist and longtime advisor to numerous world leaders, delivered a major speech at the European

Chapter 13
NATO's Expansion and the
War in Ukraine

Parliament on February 19, 2025, addressing the wars in Ukraine, Gaza, and elsewhere. He accused the United States, the EU, and NATO of decades-long warmongering and called for a radical shift in foreign policy. Sachs holds a BA, MA, and PhD from Harvard University and formerly taught there before joining Columbia. His views, though controversial, are echoed by a growing number of voices calling for multipolar diplomacy and international accountability.

A sustainable solution may require creative diplomacy to establish a new European security framework that balances NATO and Russian concerns. Simply reverting to the pre-2022 status quo is unlikely to produce lasting peace. One possible approach involves guaranteeing Ukrainian neutrality, similar to Austria or Finland during the Cold War. This would include arms control agreements, conflict-resolution mechanisms, and economic cooperation initiatives, creating a buffer zone that honors Ukrainian sovereignty while reducing NATO-Russia friction.

However, this won't be easy. Years of conflict have deepened mutual mistrust and created hardened positions on all sides. Yet the stakes could not be higher. Continued conflict increases the risk of NATO's direct involvement, which could spiral into a larger European or even global war. While the use of nuclear weapons remains unlikely, it cannot be entirely ruled out, particularly if Russia perceives an existential threat from NATO actions.

The Ukraine crisis is a cautionary tale. It shows what happens when history is ignored, diplomacy is sidelined, and power politics override empathy. Going forward, leaders must embrace a more nuanced, inclusive vision of global security one that prioritizes people over profit, peace over posturing, and cooperation over conquest. The people of Ukraine, Russia, and Europe deserve no less.

Chapter 14
Persecution & Achievements
of Jewish People

Jewish people have endured persecution for thousands of years, from 400 years of slavery in Egypt and long captivity in Babylon in ancient times to various forms of anti-Semitism worldwide, culminating in the Holocaust in the past few thousand years. Anti-Semitism persisted even after the establishment of the State of Israel. According to the Anti-Defamation League, anti-Semitic incidents in the U.S. increased from fewer than 100 in 1979 and 751 in 2013 to 3,697 in 2022.

The ADL also reported that there were "9,354 antisemitic incidents across the United States. This represents a 5% increase from the 8,873 incidents recorded in 2023, a 344% increase over the past five years and a 893% increase over the past 10 years. It is the highest number on record since ADL began tracking antisemitic incidents 46 years ago. "

Despite past persecution, Jewish people have done well. They value education, work hard, and are smart. We should celebrate their achievements and learn from them. I have Jewish friends in the U.S. and Israel. Some of them are relatives of the survivors of the Holocaust. Today, Jewish people are some of the wealthiest and most powerful groups of people on Earth. In 2023 and 2024, six of the 12 richest Americans are Jewish. According to Forbes, there were 267 Jewish billionaires with a combined net worth of 1.7 trillion dollars in 2022.

It is common knowledge that Jews dominate much of the finance industry, communication industry, entertainment industry, social media, Internet search, and many key government officials in powerful countries. Some examples of the Jewish-owned corporations that rule the world include Goldman Sachs, JP Morgan Chase Bank, BlackRock, Blackstone, DreamWorks, ViacomCBS, Facebook (Meta), Google (Alphabet), Oracle, and countless other corporations.

Powerful owners of these corporations include David Solomon, Jamie Dimon, Larry Fink, Stephen Schwarzman, Steven Spielberg, Sumner Redstone (originally Rothstein), Mark Zuckerberg, Larry Page, Sergey Brin, and Larry Ellison. I once asked Sumner Redstone who was the most influential person in his life, and he said it was his mother, who emphasized "Education, education, and education."

Key Figures and Statistics

- At year-end 2018, JPMorgan Chase Bank owned 29.5 percent of the Federal Reserve. In 2024, JPMorgan Chase Bank was the largest bank in the world by market capitalization. J.P. Morgan was instrumental in the founding of the Federal Reserve, which is a private bank. The Federal Reserve could keep the perpetual wars going in the Middle East, Ukraine, and other locations, killing many innocent people and increasing national debts that will be a burden not only to current but future generations. The Federal Reserve enabled the financing of WWI, WWII, the Korean War, the Vietnam War, the Middle East Wars, the Afghanistan War, the Ukraine War, the Israel-Gaza War, and other wars that took the lives of more than 100,000,000 people and injured many more. If and when there is a WWIII, the Federal Reserve will certainly be involved in financing the war that will wipe out human civilization. It is an existential threat to the entire world. The Federal Reserve has also enslaved billions of people to poverty and must be abolished. It is the worst financial weapon ever created against humanity. Moving forward, the world is going through major transitions in many areas. Take action while you can to save yourself and mankind.

• Digital fiat currency will be the new money. We need to make sure we understand what it is and how it will work, especially its use in surveillance where privacy is compromised.

The achievements of the Jewish people are simply remarkable and astonishing. Their accomplishments should be acknowledged.

Since 1995, Jewish individuals who have served as Treasury Secretary include Robert Rubin, Larry Summers, Jacob Lew, Steven Mnuchin, and Janet Yellen. Secretaries of State Henry Kissinger, Madeleine Albright, and Anthony Blinken are Jewish. From 1979 to 2018, Federal Reserve Chairmen Paul Volker, Alan Greenspan, Ben Bernanke, and Janet Yellen were all Jewish.

Supreme Court Justices Louis Brandeis, Ruth Bader Ginsburg, Stephen G. Breyer, and Elena Kagan are Jewish. The last three served simultaneously as one-third of the nine Supreme Court justices for a decade. Justice Elena Kagan is the first person appointed to the Court without prior judicial experience since William Rehnquist and Lewis F. Powell Jr. She is the eighth Jewish justice in the court's history.

Influential senators such as former Senate Majority Leader Chuck Schumer, Bernie Sanders, and Dianne Feinstein are all Jewish. In the current 119th U.S. Congress, 10% of senators are Jewish. Of the recipients of the Nobel Prize and the Nobel Memorial Prize in Economic Sciences between 1901 and 2023, at least 214 out of 965 individuals were Jews or people with at least one Jewish parent, representing 22% of all recipients. Most Ivy League and other top universities have Jewish presidents, professors, and significant Jewish student representation. Jewish people value education, they work hard, and they are smart. We should learn from them.

Jewish people have made substantial contributions to the U.S., the EU, and the world as a whole. I especially appreciate the Jewish doctors, nurses, scientists, engineers, lawyers, and investors who make this world a better place. On the dark side, there are also extremely wealthy Jewish people who harm society. Harvey Weinstein, Jeffrey Epstein, and Sam Bankman-Fried are some examples of despicable people.

In 2020, Weinstein was sentenced to 23 years in prison for rape in New York. He was found guilty of additional charges on December 19, 2022, which added 16 more years to his prison sentence in California. He co-founded the entertainment company Miramax, which produced several successful independent films. Harvey Weinstein, once worth $300 million, sexually abused many famous actresses and other women. In 2024, New York's top court overturned Weinstein's conviction based on legal technicalities.

Jeffrey Epstein was accused of running a vast human-trafficking operation in which he and co-conspirators procured women and girls, mostly teenagers between 14 and 17, for sex with himself and his elite associates. He was worth more than $500 million. Epstein died in 2019

109

while jailed and awaiting trial. He was continuously filmed with two cameras, but both cameras supposedly failed. There is no video of how he killed himself if he did. President Bill Clinton had a close relationship with Jeffrey Epstein. Will politicians continue to protect with impunity the rich and powerful pedophiles associated with Epstein?

Sam Bankman-Fried was convicted of fraud and related crimes in November 2023. He founded the FTX cryptocurrency exchange and, at its peak, was the 41st-richest American on the Forbes 400. Many large, illicit transactions are done with cryptocurrency, which is backed by nothing. Some call cryptocurrency the largest Ponzi scheme in the world. A Ponzi scheme is an investment fraud that pays existing investors with funds collected from new investors. FTX and FTX US had an estimated $8.7 billion combined shortfall in bankruptcy. $8.7 billion is a lot of money. Many people were deceived at a time near the end of times.

Knowing more about these despicable people would only make you angry and raise your blood pressure. So, I'll stop here.

Regarding cryptocurrency, here are 100 reasons why people should be cautious about investing.

1. High volatility and unpredictable price fluctuations
2. Lack of regulatory oversight and protection
3. Potential for fraud and scams
4. Cybersecurity risks and vulnerabilities
5. Difficulty in understanding the underlying technology
6. Absence of intrinsic value backed by tangible assets

7. Limited adoption and acceptance as a mainstream currency
8. Susceptibility to market manipulation and pump-and-dump schemes
9. Potential for significant financial losses
10. Lack of consumer protection and recourse in case of theft or fraud
11. Complexity in securely storing and managing private keys
12. Potential for hacking and theft of digital wallets
13. Absence of insurance or government-backed deposit guarantees
14. Difficulty in accurately valuing cryptocurrencies
15. Potential for regulatory crackdowns and bans
16. High transaction fees during periods of network congestion
17. Slow transaction processing times compared to traditional payment methods
18. Irreversibility of transactions and lack of chargebacks
19. Limited options for converting cryptocurrencies to fiat currencies
20. Potential for technological obsolescence and displacement by newer cryptocurrencies
21. Lack of widespread merchant acceptance of goods and services
22. Difficulty in integrating with existing financial systems and infrastructure
23. Potential for money laundering and illegal activities
24. Absence of physical form and tangibility
25. Dependence on Internet connectivity and electricity for access
26. Potential for coding errors and software vulnerabilities
27. Lack of standardization and interoperability between different cryptocurrencies

28. Difficulty in estate planning and transferring assets to heirs
29. Potential for government interference and seizure of assets
30. Lack of stability and reliability as a store of value
31. Potential for environmental damage due to high energy consumption of data mining
32. Difficulty in obtaining accurate and reliable information and advice
33. Lack of transparency in the development and governance of cryptocurrencies
34. Potential for concentration of wealth and centralization of control
35. Difficulty in using cryptocurrencies for everyday transactions and purchases
36. Potential for social engineering attacks and phishing scams
37. Lack of legal recognition and enforceability of contracts
38. Difficulty in complying with tax and reporting requirements
39. Potential for market saturation and declining returns on investment
40. Lack of stability and predictability in mining rewards and incentives
41. Potential for 51% attacks and network consensus failures
42. Difficulty in managing and securing multiple cryptocurrency holdings
43. Lack of fungibility and traceability of individual cryptocurrency units. Fungibility is the ability of a good or asset to be readily interchanged for another of like kind.
44. Potential for ideological and philosophical disagreements within communities

45. Difficulty in assessing the credibility and reliability of cryptocurrency projects
46. Lack of customer support and dispute resolution mechanisms
47. Potential for addiction and obsessive behavior related to trading and speculation
48. Difficulty in explaining and justifying investments to family and friends
49. Lack of familiarity and comfort with digital and decentralized technologies
50. Potential for social and economic disruption and instability
51. Difficulty in diversifying and managing risk in cryptocurrency portfolios
52. Lack of liquidity and market depth for some cryptocurrencies
53. Potential for pump-and-dump schemes and market manipulation by whales
54. Difficulty in assessing the long-term viability and sustainability of cryptocurrencies
55. Lack of protection against front-running
56. Potential for flash crashes and sudden market downturns
57. Difficulty in hedging and mitigating risk exposure
58. Lack of transparency in the use of proceeds from initial coin offerings (ICOs)
59. Potential for scams and fraudulent ICOs and token sales
60. Difficulty in distinguishing between legitimate and fraudulent cryptocurrency projects
61. Lack of standards and best practices for secure cryptocurrency storage and custody
62. Potential for user error and accidental loss of funds
63. Difficulty in recovering lost or forgotten passwords and private keys

64. Lack of interoperability and compatibility with existing financial infrastructure
65. Potential for market fragmentation and lack of liquidity across exchanges
66. Difficulty in assessing the impact of forks and protocol changes on investments
67. Lack of protection against market abuse
68. Potential for regulatory arbitrage and jurisdiction shopping
69. Difficulty in complying with know-your-customer (KYC) and anti-money laundering (AML) requirements
70. Lack of standardization in terminology and definitions across the industry
71. Potential for reputational damage and association with illegal activities
72. Difficulty in explaining and justifying the value proposition of cryptocurrencies
73. Lack of familiarity and comfort with cryptographic concepts and techniques
74. Potential for social and political backlash against cryptocurrencies
75. Difficulty in assessing the impact of geopolitical events on cryptocurrency markets
76. Lack of protection against state-sponsored attacks and cyber-warfare
77. Potential for concentration of mining power and centralization of control
78. Difficulty in participating in governance and decision-making processes
79. Lack of transparency in the use of funds by cryptocurrency foundations and organizations
80. Potential for conflicts of interest and self-dealing by cryptocurrency insiders

81. Difficulty in valuing and pricing cryptocurrency-based products and services
82. Lack of standardization in accounting and financial reporting practices
83. Potential for market manipulation through wash trading and spoofing
84. Difficulty in complying with securities laws and regulations
85. Lack of consumer education and awareness about cryptocurrency risks and benefits
86. Potential for scams and fraudulent cryptocurrency-based investment schemes
87. Difficulty in assessing the credibility and track record of cryptocurrency developers and entrepreneurs
88. Lack of protection against market abuse
89. Potential for price manipulation through coordinated buying and selling activities
90. Difficulty in managing and mitigating the risks of margin trading and leveraged positions
91. Lack of transparency in the operation and security of cryptocurrency exchanges
92. Potential for flash crashes and market disruptions due to automated trading algorithms
93. Difficulty in valuing and pricing cryptocurrency derivatives and other complex financial instruments
94. Lack of standardization in the classification and taxonomy of cryptocurrencies and tokens
95. Potential for scams and fraudulent cryptocurrency-based cloud mining and staking schemes
96. Difficulty in assessing the environmental and social impact of cryptocurrency mining operations
97. Lack of protection against market manipulation and front-running by high-frequency traders
98. Potential for scams and fraudulent cryptocurrency-based multi-level marketing (MLM) schemes

99. Difficulty in managing and mitigating the risks of investing in cryptocurrency ETFs and other fund products
100. Lack of a clear and compelling use case for many cryptocurrencies beyond speculation and trading.

These concerns highlight the need for caution, due diligence, and a thorough understanding of the risks involved before investing in cryptocurrencies. It's essential to approach cryptocurrency investments with a critical eye and to be prepared for the possibility of significant losses in a highly volatile and speculative market.

The United Kingdom, formerly the largest and richest empire in history, is now broke and broken, according to the UK Prime Minister Starmer's office. However, the UK holds a significant amount of U.S. debt, with $779.3 billion. The only country that holds more US debt, about $1 trillion, is Japan. The US debt is about $37 trillion as of May 22, 2025. If the UK and Japan sell large quantities of US debt, the US bond market would suffer significantly.

The UK had 7 prime ministers in the past 17 years. The number of British billionaires still stands at 165 in 2024. The US's roughly 760 billionaires now hold 3.8% of U.S. wealth, according to Americans for Tax Fairness, while the bottom half of American families control only 2.5%.

Money is manmade. We have already seen the destructive power of the Federal Reserve globally by creating money out of nothing, leading to the deaths of 100 million people and fueling the inequality gap between the super-rich and billions of people. WWI and WWII, facilitated by the Fed, turned many people away from believing in God.

Cryptocurrencies are another manmade fiat money with no intrinsic value. In both cases, the greed of a few people is very evident, and there could come a time when most people will no longer agree to live under the oppressive monetary systems that are inherently unfair, especially to young people with little or no assets.

Section 3: My Blessed and Grateful Life

Chapter 15
Early Life in Burma

Born in 1959 in Rangoon, Burma (now Yangon, Myanmar), I grew up in a Buddhist culture that taught meditation, service to others, and reverence for all life as sacred. Our household also embraced elements of Chinese folk religion, particularly ancestor worship. Burma was once the richest country in Southeast Asia, and U Thant, a distinguished Burmese diplomat, served as the third Secretary-General of the United Nations from 1961 to 1971. He held the office for a record ten years and one month, gaining global respect for his leadership. Decades later, my uncle Steven followed a similar path, working as a senior executive for the UN for seventeen years.

Watching the Apollo 11 moon landing on July 20, 1969, sparked my fascination with science. From an early age, I felt a profound connection to both science and spirituality two forces that have shaped my life ever since.

My paternal grandfather, a successful entrepreneur who owned more than a dozen businesses, had a profound influence on me. Despite his wealth, he remained humble, consistently emphasizing the value of hard work, integrity, and generosity. He often told stories laced with ancient Chinese wisdom and introduced me to traditional Chinese literature, scriptures, and proverbs. His example of quiet strength and leadership left a lasting mark on me.

Everywhere we went, he commanded respect not through fear, but through dignity. I never saw him lose his temper or fall ill, not even with a cold. His disciplined lifestyle and attention to health impressed me even as a child.

119

Chapter 15
Early Life in Burma

My paternal grandparents, Chin Lin Ngoon and Chow Toy King, showered me with love. They often prepared my favorite dishes and took me out for dim sum. My grandmother's rich chicken soup and her delicacy made from pig brains, both requiring great effort and care, made me feel especially cherished though sometimes guilty for being their favorite grandchild.

In school, I was a high achiever and skipped first grade. By fourth grade, though, my focus shifted more toward play. A childhood crush rekindled my academic ambition, and within six months, I reclaimed the top spot in class. At age 11, I began waking up at 4 or 5 a.m. to study by candlelight on the veranda.

Blessed with what I can only describe as a photographic memory, I could recall the exact words and layout of pages from books. I became an avid reader two books a week for leisure favoring genres like romance, mystery, and science fiction. Comic books also held a special place in my heart. This love for reading, which began around age 12 or 13, stayed with me through much of my teenage years.

I had many hobbies: soccer, chess, table tennis, badminton, stamp collecting, and coin collecting. A vibrant chess scene near my home drew me in, and I quickly rose in skill, often defeating seasoned players even after I moved to the United States. I once challenged a chess computer program, losing 75 times in a row before finally winning. After that, I won most of the games against it. This persistence in learning has always defined me. As long as I'm learning something new, boredom rarely finds me.

Burmese festivals and public celebrations were a joyful part of life. Certain times of the year, the streets came alive with pop-up stages for free theatrical performances. As a teenager, I loved attending these performances, even by myself, since the community environment felt safe. Thingyan, the Burmese New Year festival with origins in the 11th century, was the highlight of the year. The days-long Buddhist festival features colorful floats parading through the streets and water stations along the way. The festival's finale marks the dawn of the New Year, when government offices and businesses shut down to allow full public participation. The hallmark of Thingyan is water throwing a spirited tradition where people douse one another from head to toe. If you step outside, getting soaked is inevitable!

Kite flying was another passion, one I took seriously and pursued competitively. I spent hours on our rooftop, battling kites up to 250 yards away. My parents constantly feared I'd fall, but I walked the sloped rooftops with confidence.

The goal was to cut the strings of other kites using my own string, which was treated with a glass-coated glue mixture to make it razor-sharp. To enhance reeling speed and increase my chances of winning, I used a 10-inch spool spun with one finger while gripping both handles a skill that demanded hours of practice.

To support my hobbies, I launched a sidewalk book-lending business near my junior high school. Across from busy shops and eateries, I set up a modest lending library filled with novels and comics. Borrowers paid a small fee, and I trusted them to return the books no names, no deposits, just trust. And they always returned them. The culture of trust in Burma at that time was remarkable. Theft never even crossed my mind.

Chapter 15
Early Life in Burma

My parents were my first role models. My father worked in accounting, and my mother was a high school teacher. Though we weren't wealthy, our home was filled with love, laughter, and encouragement. I'm the eldest of four children. My parents emphasized education and excellence, sacrificing much to offer us opportunities they never had.

When I was sick as a young child, my father carried me on his back through public buses and trains to reach the children's hospital. Later, he'd take me to soccer matches and boxing events. He taught me to always have a backup plan, to expect the unexpected, and to be adaptable.

My mother was equally formative in my upbringing. We loved and respected her deeply and never dared misbehave in her presence. She never used corporal punishment but maintained discipline through calm authority. We especially loved her curry dishes. I have cherished memories of her taking me to Burma's historic capital, Mandalay.

Mandalay, Myanmar's second-largest city after Yangon, lies on the east bank of the Irrawaddy River, 631 kilometers (392 miles) north of Yangon. As of 2025, its population is approximately 1.59 million. Mandalay holds cultural significance as the final royal capital of Burma and a spiritual center for many Buddhists.

On March 28, 2025, a magnitude 7.7 earthquake struck north of Mandalay along the Sagaing Fault. The energy released was equivalent to 334 atomic bombs, making it Myanmar's strongest quake since the 7.9-magnitude temblor in Taunggyi in 1912. As of April 9, 2025, the death toll had reached up to 5,350, with 7,860 people injured and hundreds still missing. The quake caused widespread devastation: 1,700 houses, 670 monasteries, 60 schools, three bridges, and thousands of temples and pagodas were damaged. Hospitals, universities, and many historic public buildings were also impacted.

The civil war, ongoing since the 2021 military coup, has displaced over three million people, making disaster relief efforts even more challenging. The combined weight of political unrest and natural disaster has placed enormous strain on the people of Myanmar.

Chapter 16
Family Heritage and Values

My family's story is one of resilience, ambition, cultural richness, and unwavering love. On my father's side, my grandfather's success in business was tempered by his humility and wisdom. He owned over a dozen companies but remained grounded, teaching me invaluable lessons about integrity, hard work, and the quiet strength of leading by example.

My paternal grandmother often prepared my favorite dishes, showcasing her love and care through the art of traditional Chinese cooking. Together, they created a warm, nurturing environment that shaped my early years and instilled in me a deep appreciation for family, food, and tradition.

On my mother's side, my grandparents, Wong Kow and Chew Thien Lan, owned New Design, a furniture company that supplied schools and the public. My maternal grandfather emigrated from China to Burma as a young man. My grandmother was born in Burma, a second-generation Chinese Burmese with strong ties to both cultures.

Over time, my grandfather learned to speak Burmese and Hindi fluently to communicate with his Indian employees. My grandmother managed the company's finances and operations, a remarkable feat for a woman in that era. Tragically, after a military coup, soldiers seized their hard-earned business a devastating blow that shattered my grandfather's spirit and led to his early death. This injustice left a lasting mark on our family and shaped my views on resilience and perseverance in the face of political upheaval.

My mother was one of ten siblings. She never met her own grandparents, but her eldest sister told me their maternal grandparents owned a thriving shipbuilding company near Asia's busiest port at the time, as well as a rice mill and other businesses evidence of their great wealth and success. Their story added a sense of pride and mystery to my family's past, as I often imagined the bustling docks and industrious spirit of those early generations.

My mother's four brothers pursued careers as civil engineers, an architect, and hospital personnel. Her five sisters, as far as I know, dedicated themselves to managing their households. My own upbringing was more closely tied to my father's side of the family, as we lived with my paternal grandparents. Though I didn't interact much with my maternal grandparents, my mother told me that they were exceptionally intelligent and possessed a quiet dignity that resonated through their children.

Within her own family, my mother was the undisputed favorite child. She accompanied her parents every evening as they relaxed and discussed things among themselves or with guests, absorbing not only their language but also their values and dreams. My father was the favorite of his mother. I'm unsure if he held the same favor with his father. But I do know my parents devotedly cared for their own parents and supported their siblings' families whenever possible. Their sense of familial duty created a powerful example of selflessness and unity that I carry with me to this day.

My mother, Nelly Chin, was a high school teacher instructing students in Chinese and English. She attended the most prestigious university in Taiwan on a full scholarship in recognition of her excellent academics. In addition to teaching, she also tutored students privately at our home. As a young child, I remember sitting quietly by her side, observing her tutor intently, fascinated by her patience, clarity, and command of both languages.

My father, George Chin, worked in the accounting department overseeing revenue from the municipal bus system and tracking the city's gasoline reserves. In the late 1970s, he joined General Motors, which was the largest corporation in the world at that time. His transition to the private sector reflected both his adaptability and the global shifts in opportunity that marked that era.

My parents always put the needs of their four children first. They were remarkable role models, embodiments of unconditional love. As my parents grew older, my youngest sister Grace and her family took on the main responsibility of caring for them, while my siblings Catherine, Richard, and I provided support whenever we could. I had the privilege of being the firstborn child, followed by Catherine, Richard, and then Grace. Birth order may have shaped our roles, but love and shared memories bound us together.

One of my parents' greatest joys was enjoying fresh seafood at our local Chinese restaurants. They often ordered steamed whole fish and live shrimp delicacies they savored. Our extended family made it a tradition to gather for dim sum or dinner whenever we could. These times together filled us with happiness, whether dining out or enjoying home-cooked meals. Food was not just sustenance but a ritual of love and connection, a language we all understood.

As far as religion is concerned, my elementary school in Burma (Myanmar) had students from a variety of religious backgrounds. Religion was never an issue for me because I was able to get along with all the students. Buddhism was the predominant religion in Myanmar, and I volunteered to sweep the shrines of the Shwedagon, one of the world's most famous pagodas.

Furthermore, I worshipped ancestors and Chinese deities, such as Kwan Yin, and followed Confucius' teachings. Further, I attended church at Christmas with Christians, who were Caucasians. This early exposure to interfaith harmony deeply influenced my worldview, fostering a respect for diverse beliefs and traditions.

In the United States, my Uncle Frank had a successful, long career at IBM. My Uncle Steven, with an MBA from a prestigious American university, held top management positions across industries. He ended his career with 17 years in senior management at the United Nations' World Food Program in Rome. His global service inspired me to think beyond borders and consider how one life could impact millions. In total, my grandparents raised 16 children!

The values instilled by my family the importance of education, hard work, integrity, and love have been the foundation of my life's journey. Their sacrifices and unwavering support paved the way for the opportunities I've been fortunate to pursue. To this day, I carry their dreams in my heart and strive to honor them through my actions.

Chapter 17
Immigration & Early Life in America

In 1973, when I was 14, our family made the difficult decision to leave Burma to pursue better educational opportunities for us children. With just $156 (equal to $1,104 in 2024) to our name the maximum allowed by the military government we immigrated to San Francisco, seeking a better life. Those early years in America were very challenging. I didn't speak much English, and we struggled financially. Everyday life felt like a constant test of patience, resilience, and adaptation.

Fortunately, a distant relative took us in for a week. My father was able to join us nine months later. In the early 1960s, when I was young, Burma had been the wealthiest nation in Southeast Asia. But a military coup in 1962 led to Burma becoming the poorest country in the region by the time we left. What once felt like a land of promise had slowly crumbled under authoritarian rule, forcing families like ours to start over abroad.

Before my father arrived, our family risked being deported. We were only able to immigrate to the U.S. thanks to my uncle Frank, my father's brother, who sponsored us. While awaiting my father's arrival, I attended adult school at age 14 to improve my English. Despite being much younger than my classmates, I was determined to catch up.

I met Mrs. Linda Malila at the adult school, where evening classes were held at the First Chinese Baptist Church in San Francisco's Chinatown. She went above and beyond to help our family navigate the immigration process, which was crucial since we couldn't legally stay without my father present. Her compassion and commitment gave us hope during an uncertain time.

Meanwhile, I briefly attended Francisco Junior High School in the Chinatown/North Beach area, known for great Italian restaurants. I remain deeply grateful for Mrs. Malila's kindness. She became one of the first Americans to truly see me not just as an immigrant, but also as a young person trying to build a future.

To help support my family, I worked in Chinese, Japanese, or American restaurants. I'll never forget those grueling shifts in hot, cramped kitchens or the feeling of being mocked by classmates for not speaking English well. The work was exhausting, but it also taught me responsibility, discipline, and grit. Yet those early hardships instilled in me a fierce drive to succeed against all odds.

With help from dedicated teachers and mentors, I was accepted into UC Berkeley as an Alumni Scholar. The Alumni Scholarship is based on merit, not financial need. That acceptance marked a turning point it was the first tangible sign that my sacrifices and determination were paying off.

During this time, I had a newspaper delivery route in San Francisco. One morning, while delivering before dawn, a passerby said something to me. I asked, "What did you say?" He slowly enunciated: "How...are...you?" Although I could understand some English, I still struggled to keep up with native speakers' pace. Moments like that reminded me how far I still had to go, but also how far I'd already come.

As a new student at Francisco Junior High, I was initially placed in an English as a Second Language (ESL) class. I vividly remember overhearing classmates laughing and saying, "He's stupid. He can't even speak English." That hurt deeply, but it also fueled my determination to prove them wrong not just to others, but to myself.

Those words stung, especially since I had been the top student back in Burma. However, by the semester's end, I transferred out of ESL into regular English classes with American-born students. I dedicated myself to earning straight A's in every subject except a B in PE due to struggling to climb a rope. That one grade became a reminder that academic excellence alone couldn't erase the physical and emotional challenges I still faced.

Throughout junior high and high school, I worked nights and weekends in restaurants to help support my family. I also secured a summer job in a U.S. Department of Agriculture (USDA) chemistry lab. Balancing school and work was demanding, but I viewed every opportunity as a step forward.

In the summer of 1974, through a Red Cross program for recent immigrants, I met Mr. Wilhelm Schaser and Mrs. Kay Schaser. They generously sponsored me and a classmate to stay with them in their beautiful home in Eureka, California, about 300 miles north of San Francisco. The Schasers built their house on land surrounded by lush gardens, fruit trees, and towering redwood trees. It felt like stepping into a world far removed from the crowded neighborhoods of the city.

Over that stay, Bill and Kay helped us refine our English, took us hiking in beautiful redwood parks, and introduced us to friends. It was an enjoyable experience. Both were alumni of nearby Humboldt State University, and Bill also graduated from UCSF. They not only opened their home but also their hearts offering encouragement and cultural exposure we'd never experienced before.

Bill was a beloved teacher at Eureka High, leading biology, PE, and special education classes. Several of his biology students won state science competitions under his guidance. He also organized overseas exchange programs for Eureka students and parents. His dedication to education and global connection left a lasting impression on me.

I had the pleasure of meeting some of Bill's former students who went on to prestigious universities like UC Berkeley and UC Davis. While Bill taught, Kay managed their household and took the lead in building their stunning two-story home. Bill and Kay continue to inspire me with their kindness and generosity. Their example taught me that mentorship could shape not just academics, but a young person's entire outlook on life.

After establishing my Silicon Valley career, I wanted to express gratitude for all Bill and Kay had done for me. I treated them to a vacation in Hawaii, where we island-hopped and enjoyed beaches like Waikiki and Diamond Head on Oahu. We went bodysurfing and snorkeling, visited plantations, saw cultural performances, and savored luau cuisine. It was a full-circle moment sharing the fruits of my success with those who had supported my journey.

Our travels also took us to the breathtaking Hawaii and Kauai islands. For the movie Jurassic Park, the actors filmed almost all of their scenes in Kauai. On the Big Island, I had the surreal experience of standing on the warm lava flows of the Kīlauea volcano. The awe-inspiring landscapes reminded me of how far life had taken me from a poor teen to a traveler and contributor.

Our family came to the US with almost no money, but I have never felt poor in my life. I have always felt loved and all my needs were met.

At George Washington High School in San Francisco, my most influential mentor was Mrs. Ann Rhine, my English teacher. Though I enjoyed Shakespeare, my limited English was challenging. Mrs. Rhine generously stayed after school to provide extra help and encouragement. Her patience made literature accessible, even poetic, despite the language barrier.

Beyond academics, Mrs. Rhine encouraged me to get involved in student government. Despite being a latecomer at George Washington High, which had around 2,500 students, I decided to run for student council against more established and popular candidates. To my surprise and delight, I won the election and became the head of the student government. That victory was more than just a title it was validation that I belonged, and that I could lead.

Mrs. Rhine's impact on my life was immense. In my yearbook, she wrote, "To Simon - Who has been the brightest spot in my life this spring! Best wishes always. I shall never forget having you as a student. Ann Rhine, May 27, 1977." During this period, I made a habit of learning 50 new English words every day. That practice became my personal ritual a way to conquer the linguistic mountain I once thought insurmountable.

Mrs. Rhine and I kept in touch as I continued my study at UC Berkeley. Unbeknownst to me, she had been privately battling lung cancer. She passed away in 1979 without ever telling me about her illness. It was only then that I learned she had been a devout Christian. Her death was a tremendous loss for me, and I took the initiative to set up a memorial scholarship in her honor. She'll always have a special place in my heart. I wanted others to benefit from her spirit, just as I had.

In the school newspaper, I wrote: "Dear fellow GHWS graduates, our school is currently raising funds for a scholarship in memory of Mrs. Ann Rhine, our beloved teacher, who passed away during the summer of 1979. In order to show our gratitude to Mrs. Rhine, let's send in our contributions – be it a penny or an ounce of gold. Thanks for your cooperation. Please make your check payable to Ann Rhine Memorial Scholarship Fund c/o GWHS, 600 32nd Ave., S.F."

Had it not been for a negative experience during my first semester at Mission High School in San Francisco, I might never have met Mrs. Rhine. Early on at Mission High, a very tall student confronted me in the restroom and tried to rob me, demanding my wallet. Despite having no martial arts training, I fought back by imitating Bruce Lee's kung fu moves. My fear transformed into adrenaline and instinct I refused to be victimized.

The would-be thief fled, but the next day, during PE class running laps around Kezar Stadium (former home of the 49ers), someone broke into my gym locker and stole my wallet. I reported the theft, and thankfully, a Good Samaritan found my wallet discarded in a trashcan and turned it in, though the cash was gone. This incident prompted my transfer to George Washington High. That transfer became a turning point, setting me on a path toward unexpected mentorship and leadership.

At GWHS, I took a class to take pictures and make my own prints as a hobby. My favorite subject was biology, thanks to the exceptional instruction of Mr. Oscar Hollander. I considered studying biology in college to become a neurosurgeon but decided chemistry offered more potential for impactful applications. Still, I never lost my curiosity about the life sciences.

Still, my fascination with biology persisted. At UC Berkeley, I took a physiology class even though it wasn't required for my major. I scored the third highest on the final exam. The 1990 launch of the groundbreaking Human Genome Project inspired me to find a biotechnology company that would use silicon chips and AI to revolutionize medicine. The idea of merging biology with technology spoke to both my past and future a way to give back through innovation.

Chapter 18
Higher Education & Early Career

UC Berkeley was the only university I applied to, which may seem foolish in hindsight. Berkeley's chemistry program was ranked number one in the United States, if not worldwide. At the time, I simply followed my instincts. During my freshman year, I was elected president of my dorm, Putnam Hall, which housed around 200 students from freshmen to seniors.

It was a diverse and intellectually vibrant environment, which further fueled my motivation to excel both academically and socially.

Berkeley opened my eyes to a bigger world. I chose to study chemistry and chemical engineering. I was fortunate to work with brilliant professors who became lifelong mentors, like Dr. Glenn T. Seaborg, the Nobel laureate who co-discovered plutonium and advised several US presidents. Dr. Seaborg modeled scientific rigor, integrity, and concern for humanity. Working in his lab, searching for new elements, taught me the thrill of discovery and expanded my sense of life's possibilities.

Dr. Seaborg was the principal or co-discoverer of ten elements. He was the only person to have an element (seaborgium) named after him during his lifetime. He also served as UC Berkeley's second chancellor. I admired Dr. Seaborg not just for his brilliant accomplishments and the help he offered me but also for his remarkable humility. My conversations with Dr. Seaborg were always enlightening and engaging. We would often hike together between the UC Berkeley campus and the Lawrence Berkeley National Laboratory.

Along the way, he would share stories of working with Presidents John F. Kennedy and Lyndon B. Johnson. Over his career, Dr. Seaborg advised ten US presidents. He also contributed to the Manhattan Project and later championed the peaceful uses of atomic energy, emphasizing responsibility in scientific advancement.

Each year, Dr. Seaborg invited his research team to a Christmas party at his home. On one occasion, he and his wife Helen visited my family at our home. When my paternal grandfather passed away, I struggled immensely with the loss. Dr. Seaborg taught me techniques to refocus my mind, a skill that has served me well ever since. With gratitude for what Dr. Seaborg had done for me, I made a substantial donation in his honor to the College of Chemistry at UC Berkeley after I became a successful entrepreneur.

As of February 2024, UC Berkeley boasts an impressive 31 Nobel Laureates in Chemistry and 34 in Physics – more than any other university worldwide in these fields. I felt very lucky to have studied these subjects with the best scientists in the world. Another key mentor was Dr. Sumner P. Davis, a physics professor who received a prestigious teaching award in 1980.

Dr. Davis had a true gift for teaching. Over his career, he supervised 36 doctoral students. In 1993, he was awarded the Berkeley Citation. He was also an accomplished glider pilot, and I had the pleasure of flying with him in his glider. Like Mrs. Rhine, Dr. Davis was a devout Christian. His lectures combined clarity, humor, and profound insight, making even the most complex physical principles feel approachable.

Dr. Davis and I continued our friendship well beyond my undergraduate years, meeting for dinner several times yearly until his passing in his eighties. When he died, I made a substantial donation to UC Berkeley in his memory to fund upgrades to his beloved Physics 111 lab. Early in my career, about two years after graduating from Berkeley, I faced a major career decision and turned to Dr. Davis for invaluable guidance.

His advice helped me navigate not only that professional crossroads but also instilled in me a habit of seeking mentors who embody both intellectual excellence and moral integrity.

When I began my education at UC Berkeley, I did not believe in Christianity. My roommate and some other people tried to witness their faith in Jesus Christ. I could not believe what they believed. Christianity seemed impossible from a scientific perspective. It is only in retrospect that I could appreciate what they did. Their sincerity and quiet confidence left a lasting impression, even if I wasn't ready to accept their beliefs at the time.

The summer after my freshman year, I took on a challenging role as a door-to-door encyclopedia and children's book salesman with the Southwestern Company based in Nashville. This experience was formative, with opportunities for learning, travel, and personal connections. I was partnered with two fellow UC Berkeley students and was sent to Norwich, Connecticut, after training.

Upon arriving in Norwich, the three of us had only $100 each and no lodging plan. As we pondered our next move over coffee at a Dunkin' Donuts, a UC Berkeley alumnus spotted my Cal cap and struck up a conversation. Learning of our plight, he generously hosted us for the night.

The kind employees, moved by our situation, slipped us scratch cards to win free donuts. We were touched by the kindness of these strangers helping us. This early experience with generosity from strangers taught me that acts of kindness can have an outsized impact, especially during times of uncertainty.

The next day, we rented a room from Reverend Neil Chadwick of the local Assembly of God church.

Reverend Chadwick and his family welcomed the three of us into their home to share a single room on the floor. We remain grateful for their hospitality. That summer was transformative and full of valuable lessons. Selling books door-to-door taught me perseverance, emotional intelligence, and the importance of connecting with people from all walks of life.

Wing Hsieh, one of my roommates, then became a lifelong friend. After studying optometry/ophthalmology in Indiana, where he met his wife, Sharon, Wing returned to the Bay Area. Sharon became my optometrist, and when their daughters went to college, I was happy to offer guidance. Wing and his family are devout Christians.

After some success selling books, I decided to move on to more fulfilling pursuits.

That summer not only provided personal growth but also introduced me to lifelong friendships grounded in trust, faith, and shared challenges.

My next stop was Raleigh, North Carolina, to visit my Uncle Frank and his family. Together, we enjoyed a beach vacation in Myrtle Beach, South Carolina. Uncle Frank, an IBM staff engineer, was based in the renowned Research

Triangle Park. Over the years, he and his family had lived in New York City, Hyde Park, Poughkeepsie, and even owned a house on land once owned by President Franklin D. Roosevelt.

Uncle Frank was an exemplary role model from whom I learned greatly.

He exemplified a blend of technical brilliance, humility, and familial devotion that left a deep imprint on my aspirations. His encouragement reminded me that success was not just about knowledge but also about integrity, perseverance, and compassion.

After visiting family, I embarked on a two-week cross-country Amtrak rail adventure with an unlimited travel pass. The 15,000-mile journey took me to Boston (with stops at Harvard and MIT), New York City, Washington D.C., Niagara Falls, Chicago, and many places in between.

I spent most nights on the train to save money and have more time to explore. My return route went through Seattle. It was an unforgettable experience. Traveling solo across the country at such a formative age broadened my horizons and reinforced a sense of independence, curiosity, and appreciation for the diversity of American life.

During my junior year at UC Berkeley, I secured an engineering internship with US Borax in Boron, California home to the largest borax mine and biggest open-pit mine in the state. My task was to create a comprehensive manual detailing the computer program simulating the borax production process, starting with raw ore from the mine.

The resulting manual was an inch thick, with diagrams illustrating chemical streams processed through towering 3-

story machines. The project was challenging yet gratifying, providing invaluable learning opportunities. The work helped bridge my classroom knowledge with real-world industrial processes and gave me hands-on experience in both chemical engineering and technical communication.

The Boron open-pit mine could be punishing, with temperatures sometimes reaching 110°F (43°C). Interestingly, the mine's waste products, discarded as low-value then, are now recognized as a precious lithium source for batteries powering devices like phones, computers, and electric vehicles. At the time, none of us interns could have imagined how the so-called "tailings" would later become critical in the global transition to clean energy.

Boron sits on the edge of the Mojave Desert, just 29 miles from Edwards Air Force Base. During my time at US Borax, I had the extraordinary luck of witnessing the Space Shuttle Columbia coming down to land at Edwards on April 14, 1981, after its inaugural mission.

Perched atop a chemical silo with colleagues, I watched in awe as the shuttle glided effortlessly through the blue sky. It was a historic moment Columbia's landing marked the successful completion of the first-ever orbital mission of NASA's Space Shuttle program, STS-1.

While at US Borax, I lived in California City, located about 65 miles from Death Valley National Park and 29 miles from Edwards Air Force Base. While camping in the park, I went into a restaurant to eat dinner. They said, "Sir, you cannot eat here because you are not wearing a suit and a tie." Who brings a suit and a tie when they are camping in a national park? It was a humorous but jarring reminder of the lingering formality in some dining establishments of the era even in the middle of the desert.

That summer at U.S. Borax, I formed a close bond with a music student from UC Santa Barbara. We spent nearly every day together playing tennis in the evenings, studying the Bible, dining out, and attending church services on Sundays. She was beautiful, athletic, and had deeper scriptural knowledge than I did.

As our connection deepened, she asked me directly, "You are in love with me, aren't you?" In a moment of hesitation that I now recognize as one of my greatest missteps, I failed to express the depth of my feelings for her. Looking back, I see that my emotional reserve masked an inner struggle between personal longing and unresolved grief.

My reticence may have stemmed from the unresolved guilt and turmoil I experienced when parting with my paternal grandparents upon leaving Burma. Their unwavering love and generosity left a deep impact. I was sorry for not having the opportunity to thank them and take care of them when old age and illness challenged their daily lives. Despite the physical distance, their presence in my heart remained constant.

I was devastated by the news of my paternal grandfather's passing just weeks after returning to UC Berkeley from that internship. My world was thrown into chaos. At the time, my father had already transferred from San Leandro to Sparks, Nevada, for his job with General Motors. With three children in college and one in private high school, our family faced significant financial strain.

At one point, I had only $2 for personal spending in an entire month, subsisting on prepaid dorm meals with sheer determination. It was a humbling time that taught me how to stretch limited resources while staying focused on my goals.

In addition to my studies, I felt a responsibility to support my family and provide academic guidance to my sister, Grace. The emotional weight of my grandfather's death impacted my academic performance.

I was extremely close to my grandfather. Three or four months later, my grandmother joined my grandfather in eternal rest. They had lived long lives, reaching ages 84 and 80, respectively, far surpassing the average mid-50s life expectancy in Burma at that time.

Both were born in China but immigrated to Burma as adults. They went back to China and built a nice home for themselves but ultimately left it for their relatives. Their selflessness and resilience are part of the legacy that shaped my values and aspirations.

My paternal grandfather rarely spoke of his business ventures. However, his brother once shared that he had built and managed between 14 and 17 thriving enterprises. Tragically, during civil unrest, rioters looted and burned down both his businesses and the family homes.

As a wedding gift to their eldest daughter, my grandparents presented her with a jewelry box the size of a cigar case, filled with 24-karat gold, jade, rubies, and other precious stones. She went on to live into her 90s. Despite losing so much, my grandparents never lost their generosity or grace.

My paternal grandparents showered me with affection. I was their favorite among their dozens of grandchildren. Losing them, compounded by the end of my relationship with the girl in Southern California, nearly overwhelmed me. In my final two years at UC Berkeley, I devoted more time and energy to studying the Bible than all my academic coursework combined.

To this day, I consider that time and spiritual dedication to be the wisest choice I've ever made. It became the cornerstone of my transformation from loss and uncertainty to purpose and faith.

Chapter 19
Silicon Valley Career &
Professional Development

After graduating from UC Berkeley with a chemical engineering degree, I accepted a position at Master Images, a premier semiconductor photomask production company. My trajectory there was swift – starting as an engineer, I was promoted to a managerial role overseeing manufacturing within just 9 months, doubling my salary in my first year of employment. Assuming responsibility for the company's largest operation 6 months later, we experienced a period of great success and rapid growth.

Photomasks are precision tools used to transfer intricate circuit patterns onto semiconductor wafers through a process known as photolithography. They are essential for fabricating integrated circuits, making photomask technology one of the most crucial yet unsung components of the global electronics supply chain.

Then, a customer I had met once on my very first day offered me a new job opportunity a 20% pay raise and stock options. Seeking advice, I turned to Dr. Davis. His guidance was simple yet profound. He asked me one question: "Which group of people would you rather work with?"

By that time, I had already recruited several UC Berkeley classmates to join the company in their first jobs after graduation. I decided to stay put. That single piece of advice underscored a lasting truth: workplace culture and trusted colleagues often outweigh short-term financial gain.

As market demand surged, we needed to double our output without expanding equipment. This meant frequent overtime shifts for production teams, even as we trained new hires. Needing to allocate experienced personnel to guide and mentor newcomers, resulting in even longer hours for seasoned workers, compounded the strain on existing staff.

In retrospect, our lack of proactive planning for increased staffing and equipment left us ill-equipped to meet the explosive growth once the economy kicked into high gear. While productive for the bottom line, this boom time placed an immense burden on operations staff.

This period coincided with the broader expansion of Silicon Valley during the early to mid-1980s, driven by growing demand for personal computers, telecommunications equipment, and consumer electronics all of which depended on semiconductor innovation.

We ultimately reached a breaking point where we could no longer guarantee on-time delivery to our marketing division. This ignited tensions between the operations and marketing vice presidents, culminating in an organizational restructuring. It was a powerful lesson on understanding business intricacies. The experience taught me that technical excellence must be matched by strategic communication, scalable infrastructure, and cross-functional alignment in order to sustain growth.

My career centered on developing semiconductor photomask production technologies and managing people from 1983 to the 1990s. This expertise led to opportunities to transfer these technologies and the global ISO 9001 quality management systems to semiconductor photomask manufacturing sites in Germany, France, South Korea, and China.

The ISO 9001 standard established in 1987 by the International Organization for Standardization was rapidly adopted in the semiconductor industry to ensure consistent product quality and global interoperability, especially critical as supply chains became increasingly international.

The terms "semiconductor," "chip," and "microchip" are often used interchangeably to describe the tiny yet powerful electronic components that are the bedrock of modern devices from smartphones and computers to AI systems, vehicle electronics, and more.

At their core, these chips are composed of layers of semiconductor material typically silicon patterned and doped to manipulate electrical conductivity. Photomasks play a central role in this precision patterning, serving as the blueprint for every microcircuit layer.

As I advanced my career, DuPont, the eighth-largest company in the US in the 1980s and the largest chemical company in the world at the time, acquired Master Images. I was on DuPont's core team to implement quality systems in Silicon Valley that earned the prestigious Malcolm Baldridge National Quality Award, which was established by the U.S. Congress in 1987 to raise awareness of quality management and recognize U.S. companies that have implemented successful quality management systems.

This recognition placed us among an elite group of American companies including Motorola and Federal Express that led a national movement toward continuous improvement and operational excellence.

I simultaneously pursued an MBA at Santa Clara University while working full-time juggling employment's demands with graduate studies' rigors. I actually enjoyed the synergy of applying what I learned in school to my work and what I learned at work to my school projects almost on a daily basis.

The Jesuit values at Santa Clara emphasized ethical leadership, innovation with integrity, and the holistic development of both intellect and character. These principles profoundly shaped my approach to business and management.

Over the years, I successfully led numerous technology transfer initiatives for DuPont, including one complex project spanning 10 sites across 7 countries. My efforts were recognized with two prestigious DuPont Achievement Awards, reflecting the impact and value of my contributions. These awards celebrated not just technical implementation but also leadership in multicultural coordination and sustainable systems development.

Early in my semiconductor photomask career, I was elected to the board of directors for our industry's trade association. In the early 1980s, I was invited to present at the annual Bay Area Chrome Users Society (BACUS) conference on using personal computers to enhance control and optimization of photomask manufacturing processes.

BACUS, originally an acronym for "Bay Area Chrome Users Society," became a global SPIE-sponsored community of experts focused on advancing photomask technology. Presenting there was a rare honor and confirmed my contributions were at the forefront of the field.

The computer program I developed for this purpose analyzed and monitored over 6,000 distinct variables, requiring a staggering 6 hours daily to execute on the era's limited IBM PC platform. Despite computational constraints then, implementing this software solution significantly improved both employee performance and overall manufacturing yield.

The program I wrote was the first in the world to collect, analyze, and store all relevant machine, chemical, and human data to timely control the effectiveness and efficiency of semiconductor photomask manufacturing. This software pioneered data-driven process control well before the term "big data" entered mainstream business vocabulary, demonstrating how early digital tools could unlock measurable gains in precision manufacturing.

I enjoyed my graduate studies and the environment at Santa Clara University. It was completely different from UC Berkeley. To me, it was like having the best of both worlds. At Berkeley, you get to meet and learn from the best professors and students from around the world every day. The conversations were stimulating.

At Santa Clara, the emphasis was on competence, conscience, and compassion. You grow more as a person and not only in academic excellence. Even former rivals from UC Berkeley and Stanford feel like a part of the Santa Clara family. This duality gave me a rare and invaluable perspective: rigorous intellectual formation from Berkeley and humanistic, ethics-centered development from Santa Clara.

The knowledge and skills gained through my MBA program, coupled with the invaluable hands-on experience amassed in the semiconductor industry, laid the groundwork for my eventual transition into entrepreneurship.

Chapter 20
International Experiences and Cultural Insights

In 1986, my professional journey took me to Stuttgart, Germany, where I was entrusted with transferring semiconductor photomask technologies to Bosch a multinational engineering and technology conglomerate headquartered in Stuttgart.

Founded in 1886 by Robert Bosch, the company has grown into a global leader in mobility solutions, industrial technology, energy and building technology, and consumer goods. As of 2023, Bosch reported €91.6 billion in revenue and remains the world's largest automotive supplier. Prestigious Bosch clients in Stuttgart include legendary automakers Mercedes-Benz and Porsche, both headquartered nearby.

Overall, my first trip to Germany and Europe included a memorable Lufthansa Airlines flight. At check-in in San Francisco, the attendant apologetically explained they had overbooked business class. Bracing for a downgrade, I was delightfully surprised when she offered, "Would you mind if we upgraded you to First Class?" I gratefully accepted, immersing myself in Lufthansa's premier cabin luxury and comfort. Lufthansa, founded in 1953, is Germany's flagship airline and known for its excellent transatlantic service.

During my time at Bosch, I witnessed the epitome of German efficiency firsthand. I thoroughly enjoyed Germany and found the people warm and welcoming a contrast to the often-stereotyped perception of Germans as cold or overly rigid.

In France, my role involved transferring cutting-edge technologies to DuPont's facility in Rousset. From June to September 1989, I had the privilege of residing in the picturesque city of Aix-en-Provence, nestled near the French Riviera. A 1989 French magazine survey ranked Aix as one of the most desirable cities to live in, praised for its architecture, Mediterranean climate, and cultural vibrancy. The vibrant university town of Aix is strategically located just 40 km from Marseille, 90 km from Avignon, 101 km from Saint-Tropez, 150 km from Cannes, and 175 km from Nice. I enjoyed visiting all of these cities.

During my stay, I was provided with a charming 2-bedroom apartment, merely a half-block from the renowned Cours Mirabeau the city's iconic main thoroughfare and cultural hub. My time in Aix was an absolute delight, filled with memorable weekend excursions across Europe including Rome, Geneva, Monte Carlo, Cannes, Avignon, Paris, and the sacred pilgrimage site of Lourdes.

I once stayed up dining with a French acquaintance until 3 a.m., reveling in the region's joie de vivre.

DuPont generously covered all my living expenses apartment, car, fuel, dining out or home cooking (though I admittedly ate out more often), international phone calls, and laundry services while my regular U.S. paycheck continued uninterrupted. My role was transferring advanced photomask production using state-of-the-art laser lithography techniques within an ultra-clean ISO Class 1 cleanroom environment, where even a speck of dust cannot be tolerated.

Maintaining an ISO Class 1 cleanroom requires the air to be meticulously filtered with fewer than 12 particles measuring 0.3 microns or smaller per cubic meter. For comparison, the average human hair is around 70–100 microns in diameter. Prior to my France assignment, I had been deeply involved in developing cutting-edge technologies in Silicon Valley to optimize the performance of photomasks produced using the first Ateq laser system deployed in a production setting. My expertise also extended to collaborating directly with the laser system manufacturer to fine-tune its capabilities for the world's first semiconductor photomask laser lithography system in production.

In transferring these technologies to the DuPont facility in France, I worked closely with the company's technical staff, operations personnel, and machine operators, often conducting meetings almost entirely in French to facilitate clear communication. This was no small feat, given the technical nature of the discussions.

Additionally, I assumed the role of project leader, overseeing the successful technology transfer implementation. The photomasks produced during the very first production run under my guidance were flawless, exhibiting zero defects a remarkable achievement that earned me a DuPont Achievement Award.

Life on the French Riviera was an absolute dream. I reveled in the breathtaking beaches, indulged in the delectable cuisine, and forged meaningful connections with the people. Within walking distance of my Aix apartment, I had over 450 restaurants to choose from. Dining on French delicacies three times daily was heavenly, yet surprisingly, I didn't gain a pound. This may have been due to the European lifestyle of frequent walking, balanced meals, and smaller portion sizes.

The most exquisite meals and finest wines were those shared in the homes of colleagues, prepared and served lovingly by their spouses. My fondness for France and the French only deepened as I witnessed their remarkable efficiency and zest for life. During these intimate home dinners, I conversed almost exclusively in French with hosts and their partners, immersing myself in the language and strengthening our bonds of friendship.

In 1989, France celebrated the 200th anniversary of the momentous French Revolution. I had the good fortune of being in Nice during the historic festivities unfolding along the iconic Promenade des Anglais seafront. For miles, the air was filled with pulsating music rhythms, including my all-time favorite band, the Eagles' "Hotel California." This song, though American in origin, resonated with audiences worldwide and added a surreal quality to the celebration.

Later that year, I witnessed Paris's spectacular Eiffel Tower centennial celebrations. The Eiffel Tower, inaugurated in 1889 during the Exposition Universelle, had become not only an engineering marvel but also a national symbol. The view from atop this marvel was breathtaking, and I fell in love with the city's timeless charm and romance.

The Louvre Museum is widely regarded as perhaps the greatest museum in the world a veritable treasure trove of art and history. As I wandered its hallowed halls for the first time, I found myself silently uttering "Oh, my God!" in awe and reverence repeatedly. Like countless others, I had seen reproductions of Leonardo da Vinci's enigmatic Mona Lisa before. But it wasn't until I stood before the original painting, encased behind bulletproof glass in a climate-controlled enclosure, that I truly grasped the depths of its allure and enduring fame. That revelatory moment left an indelible mark, and I knew I would gladly journey back to Paris time and again just to bask in the presence of this singular masterpiece.

In Paris, I also enjoyed the Musée d'Orsay and other museums. Located on the Left Bank of the River Seine, this former Beaux-Arts railway station originally built for the 1900 Exposition Universelle now houses the world's most comprehensive collection of Impressionist and Post-Impressionist masterpieces.

It features works by Berthe Morisot, Claude Monet, Édouard Manet, Degas, Renoir, Cézanne, Seurat, Sisley, Gauguin, and van Gogh. It is one of the largest art museums in Europe. According to The Art Newspaper's annual visitor figures, in 2022, Musée d'Orsay welcomed 3.2 million visitors, making it the sixth-most-visited art museum globally and the second-most-visited in France, after the Louvre.

Visiting the Notre-Dame Cathedral was quite memorable, inside and out. I also spent more than 15 minutes on its roof enjoying the breathtaking panoramic view of Paris. The cathedral, an iconic masterpiece of French Gothic architecture completed in the 14th century, suffered a devastating fire in 2019. Notre-Dame Cathedral officially reopened to the public on December 8, 2024, after a five-year restoration.

The technologies I brought to South Korea played a pivotal role in bolstering that nation's semiconductor photomask industry, and I take pride in contributing in my own small way to South Korea's technological advancement. South Korea's rise as a semiconductor powerhouse led by companies like Samsung and SK Hynix has been crucial to its transformation into one of the most advanced digital economies in the world.

Similarly, my work transferring cutting-edge semiconductor photomask manufacturing techniques some I had personally developed to facilities in Wuxi and Shanghai during the late 1980s and mid-1990s was instrumental in laying the foundation for China's remarkable rise on the global stage.

As I witness China's astonishing progress in lifting 800 million citizens out of poverty between 1981 and 2015, I am filled with gratification knowing my efforts played a part in this transformative journey. When I left Burma, my grandfather said, "China would be a strong country in 30 years." Decades later, his words proved remarkably prescient.

I went to Japan for business up to six times a year, visiting leading semiconductor companies all around the country. Doing business in Japan is completely different from doing business in the United States, requiring different levels of preparation. Japanese business culture emphasizes respect, hierarchy, punctuality, and consensus-driven decision-making.

Tokyo was very expensive. In the 1990s, a hotel room cost more than $300 per night. Dinner could cost hundreds of dollars. This was during the peak of Japan's economic bubble, when Tokyo was considered one of the most expensive cities in the world. Japanese people value freshness in seafood and are willing to pay for it. The price of shrimp on ice in the afternoon is half the price in the morning.

Covering 230,000 square meters in central Tokyo, Tsukiji Fish Market was a major tourist attraction for both domestic and overseas visitors. It was fascinating to visit this largest wholesale fish and seafood market in the world until its closure in 2018.

The inner wholesale market was then relocated to Toyosu Market, which opened in October 2018 and is nearly twice the size, offering high-tech sanitation and temperature control. I loved eating sushi both in Japan and Korea, where I especially liked grilled live eel (unagi kabayaki). Also, the efficiency and on-time arrivals of Japanese Shinkansen bullet trains capable of speeds exceeding 300 km/h (186 mph) were simply remarkable, setting the global standard for punctuality and safety.

In the late 90s, I went to Korea to visit Samsung and other semiconductor companies. I always enjoyed staying at the Grand InterContinental Seoul Parnas. The generously sized suites came with separate living rooms at a great price, far less than Tokyo, and much more comfortable. It was exciting to see Samsung's flagship product dynamic random-access memory (DRAM) chips in development three years ahead of market introduction, underscoring their role in shaping global tech trends.

These international experiences not only enhanced my professional skills but also broadened my cultural understanding and global perspective. Each country presented unique challenges and opportunities, requiring adaptability and fostering a deep appreciation for diverse work cultures and practices. Working across Asia and Europe helped me evolve both technically and personally enriching not only my career, but also my entire outlook on life.

Chapter 21
Entrepreneurship in Biotechnology and Artificial Intelligence

The groundbreaking Human Genome Project, an international scientific effort to map the DNA of certain organisms, stands as one of the most ambitious undertakings in human history. The project's successful mapping of the human genome, launched in 1990 and officially completed in April 2003, inspired me to establish Iris Biotechnologies Inc. in 1999. The company's focus was on developing innovative technologies to enhance personalized breast cancer diagnosis and treatment by analyzing genetic, medical, and lifestyle information.

Over the years, I've had the privilege of collaborating with patients and physicians at esteemed institutions such as Stanford University, University of California San Francisco (UCSF), Sutter Health, Kaiser Permanente, St. Joseph Health, and Dignity Health, among others.

I've been fortunate to have scientific advisors from top institutions like UCSF, MD Anderson Cancer Center, Baylor College of Medicine, Wake Forest University, and the University of Hong Kong. To facilitate these collaborations, Iris Biotechnologies opened offices across California in Santa Clara (near Stanford University), San Leandro (close to UCSF and UC Berkeley), and Davis (adjacent to UC Davis).

To build the Iris Biotechnologies team, I recruited PhDs and MDs from elite universities and medical centers, including UCSF, Stanford, UC Berkeley, UC Davis, MD Anderson, and Parke-Davis (later acquired by Pfizer). My investors and board members include graduates from

156

prestigious universities such as Harvard, Yale, UCSF, UC Berkeley, and Stanford. Two prominent U.S. law firms also became equity investors in Iris Biotechnologies, a relatively rare endorsement in the life sciences sector.

Though offered venture capital financing, I chose to decline. Instead, we focused on filing multiple patent applications across major global jurisdictions, including the United States, European Union, Japan, Canada, China, Australia, and New Zealand. Every single one of our patent applications was ultimately granted. One of our core patents titled "Artificial Intelligence System for Genetic Analysis" outlined a platform for integrating genomic, medical, and lifestyle data to enable truly personalized medicine. I personally traveled to the European Patent Office in Munich, Germany, to present our case, where we successfully secured protection.

Both the French and UK governments invited me to meet with key business leaders, academics, clinicians, and officials to discuss potential international expansion. In France, I visited Paris, Lyon, Grenoble, Provence, and other regions to tour facilities, hear incentive offers, and explore business opportunities.

In the UK, I went to London, Oxford, and Cambridge for meetings with clinicians, professors, and government officials. During one visit, I was issued a special permit allowing me to bypass standard immigration lines an uncommon courtesy and a sign of strong diplomatic and business interest.

I also traveled to China to investigate possibilities for medical collaboration. I delivered talks at the "Medicine in the 21st Century" international symposium for doctors, scientists, and technologists, as well as other medical conferences. My visits took me to Shanghai, Beijing, Tianjin, Dalian, Chengdu, Hong Kong, and other cities. At the time, China was rapidly expanding its investment in biotechnology and genomics, and my work aligned with their national push toward personalized medicine.

Within the US, I had the honor of being invited to speak at the BIO International Convention the world's largest biotechnology conference and exhibition, drawing over 15,000 professionals from more than 4,000 organizations across 65 countries.

As a recognized expert holding three patents on artificial intelligence systems for genetic analysis, I was invited to present at conferences like the American Association for Clinical Chemistry (AACC) Annual Meeting the largest global gathering of laboratory medicine professionals and the Global Wellness Summit, where leaders shape the future of the global wellness industry through science, business, and holistic health innovation.

In spring 2008, Frost & Sullivan honored Iris Biotechnologies with their North American Technology Innovation Award in Pharmacogenomics, noting: "For patients, this technology would be pivotal in yielding a personalized treatment regimen, with the greatest possibility of success, for each individual patient's particular disease like cancer, heart disease, diabetes, or gene-related metabolic disorders."

My typical workweek usually involved 15–16 hour days during the week and a few additional hours on weekends. But I absolutely love what I do.

In July 2008, PharmaVOICE magazine recognized me as one of the 100 most inspiring people in the life sciences industry. They wrote, "These innovative, forward-thinking executives (I was one of twelve entrepreneurs) have not only carved out a niche in the life sciences, in many cases creating a whole new business area, but have also successfully steered their companies to new heights."

When Iris Biotechnologies began public trading on the stock market in August 2008 (OTCBB: IRSB), I gifted stock to my siblings, their spouses, and their children the maximum amount permitted by law under the annual gift tax exclusion, which was $12,000 per recipient at that time.

In 2010, after the National Institutes of Health (NIH) reviewed Iris Biotechnologies' technology, the government awarded us a $250,000 grant to further our work in personalized and targeted medicine. This marked important federal validation of our platform's potential in real-world clinical applications.

I've been privileged to have conversations with visionary scientists like Dr. James Watson, co-discoverer of the DNA double helix structure in 1953 and first director of the Human Genome Project, and Dr. Thomas Cech, Nobel Laureate in Chemistry and former president of the Howard Hughes Medical Institute. They shared insights into the future of scientific research and medicine. Their perspectives helped guide my strategic decisions, particularly in navigating the intersection of science, regulation, and patient care.

When I asked Dr. James Watson in April 2003, at a celebration marking the 50th anniversary of the discovery of the DNA double helix, how long he thought it would take for human genome information to be widely used in medicine, he predicted it would take at least 25 years. His estimate has proven prescient, as we continue to bridge the gap between genomic data and real-world medical application today. I found my conversation with Dr. Thomas Cech, over lunch at Howard Hughes Medical Institute (HHMI) headquarters in Chevy Chase, Maryland, equally fascinating.

In 2014, I founded Iris Wellness Labs with the goal of providing in-depth scientific analysis to offer insights into complex medical conditions like cancer, cardiovascular disease, immune disorders, diabetes, and obesity for patients and doctors at leading health centers. Our focus was on integrating multi-omics data genomics, microbiomics, and metabolomics with advanced analytics to support precision and preventive healthcare.

I am grateful to Dr. Douglas Hendren and my sister, Grace Osborne, for their roles as board members of both Iris Biotechnologies and Iris Wellness Labs. Dr. Hendren was educated at Harvard University and Case Western Reserve University. Grace was educated at UC Davis. They both also have MBAs and bring multidisciplinary insights to our mission of personalized medicine. I am thankful to Dr. Daniel Farnum, educated at UCSF, for guiding Iris as a board member for more than a decade.

I also want to acknowledge Mr. Wilhelm Schaser and Mrs. Kay Schaser for the many things they have done for Iris, including their steadfast support and involvement in operational and strategic initiatives.

The level of detailed knowledge Iris has in analyzing individual genomes, gut microbiota, and nearly 900 key blood metabolites is not yet standard practice in most US hospitals today. However, the rise of large-scale data platforms and artificial intelligence (AI) could make such in-depth analysis a part of standard medical care worldwide going forward.

In the decades following the groundbreaking 1953 discovery of DNA's double helix structure by James Watson and Francis Crick, the prevailing wisdom in science and medicine held that specific genes play a key determining role in shaping the characteristics that define us as humans. While mapping the human genome has yielded many insights into our genetic makeup, it offers only a partial blueprint of our biological complexity.

First of all, DNA by itself cannot produce anything. The human egg cell, where the sperm and egg are fertilized, plays a crucial role in the propagation of life. DNA alone is useless without all the other cellular organelles, molecular machinery, and nutrients within the egg. DNA needs the ecosystem inside an egg to thrive. It is within this microenvironment that the first molecular signals initiate cell division and development.

Only about 1.5% of the human genome is actually dedicated to encoding proteins the fundamental building blocks of life. The remaining 98.5%, previously called "junk DNA," houses non-coding regulatory elements, introns, and sequences crucial for gene expression, chromatin structure, and epigenetic regulation.

Throughout an individual's lifespan, an intricate interplay of gene expression unfolds, with different genes being activated or silenced at specific points to orchestrate the processes of growth, development, and sustenance. However, genes alone don't paint a complete picture of human biology's complexities.

Nested within each cell lies a dynamic network of protein signaling pathways that profoundly influence gene behavior through epigenetic mechanisms chemical tags such as DNA methylation and histone modification that regulate gene activity without altering the underlying DNA sequence. This allows the cell to adapt to internal and external cues.

Beyond genetics and epigenetics, the human microbiome the vast ecosystem of microorganisms residing within and upon our bodies and the countless bioactive metabolites they produce also play a crucial role in shaping our overall health and well-being. Recent research shows that microbiome diversity can impact immune function, mood, and even response to certain therapies.

Achieving optimal health is multi-faceted, extending beyond our genetic blueprint to encompass lifestyle factors like adequate sleep, regular exercise, a balanced nutritious diet, and cultivating a life full of love, joy, and fulfillment. A baby has no choice in what it eats or drinks, where it sleeps, or where it can go. It is entirely dependent on others for nurturing. As we grow, we generally have more choices in what we do with our lives. As humans in a modern society, we are interconnected and interdependent, relying on one another to thrive in a complex and rapidly changing world.

The advent of CRISPR-Cas9, a revolutionary gene-editing tool discovered in 2012, represents a quantum leap in our ability to manipulate life's very building blocks. In essence, it functions as a molecular scalpel or a word processor for the genetic code, allowing targeted changes to DNA sequences with unparalleled precision and ease.

The scientific community long anticipated that Dr. Jennifer Doudna, UC Berkeley's Li Ka Shing Chancellor's Chair in

Biomedical and Health Sciences, and Dr. Emmanuelle Charpentier would be recognized with the Nobel Prize in Chemistry, which they indeed received in 2020 for their groundbreaking CRISPR-Cas9 contributions.

Dr. Charpentier, a luminary in her own right, is a French professor and researcher specializing in microbiology, genetics, and biochemistry. Since 2015, she has directed the Max Planck Institute for Infection Biology in Berlin.

The implications of CRISPR-Cas9 are nothing short of staggering, with far-reaching applications spanning genome engineering, disease modeling, cancer therapy, infectious disease control, and even agriculture and environmental science.

Chapter 22
Personal Growth, Giving Back, and Reflections

My passion for aviation began in my freshman year at UC Berkeley, where I took a class on preparing students for the written private pilot's license exam. Our instructor was a former Navy fighter pilot who had flown the legendary McDonnell Douglas F-4 Phantom II, a tandem two-seat, twin-engine, long-range supersonic jet interceptor and fighter-bomber used extensively during the Vietnam War. From the moment I first took to the skies, I was hooked on flying's unparalleled sense of freedom and exhilaration.

As my graduation from UC Berkeley with a chemical engineering degree approached, I met with the US Air Force recruiter to explore becoming a fighter pilot. The recruiter expressed enthusiasm about me joining as a navigator but regretfully informed me that my need for corrective lenses disqualified me from being a pilot. This was devastating, as my eyesight had been perfect when I started college but deteriorated to needing glasses by graduation. At that time, Air Force regulations required 20/20 uncorrected vision for pilot candidates, a requirement that has since been revised for some aviation roles.

Undeterred, I resolved to complete pilot training and get my private license after I got established in my professional career. I decided to become a pilot, even if not a military one.

I strongly believe in the importance of giving back to society. One way I've done this is by serving on the inaugural advisory board of Santa Clara University's Ignatian Center, an esteemed center that positively impacts many lives. The vision of the Center is to provide leadership for the integration of faith, justice, and intellectual life by embracing Jesuit wisdom to inspire awareness, thought, reflection, discernment, and action.

The Center's mission is to focus on a holistic approach to learning, a commitment to open and inclusive dialogue, and finding God's will for better decision-making. Founded in 2005, the Ignatian Center for Jesuit Education fosters community-based learning and global engagement aligned with Jesuit values. The university's president, Fr. Paul Locatelli, S.J., invited me to join the board, and I was honored to serve for eleven years. I also made significant donations, six figures, to Santa Clara University.

Additionally, I served for nine years on the Industrial Advisory Board of the University of the Pacific's Thomas J. Long School of Pharmacy and Health Sciences, as well as six years on the board of directors for the YMCA of Silicon Valley. Our YMCA board held fiduciary responsibility for $70 million while serving around one million members annually across multiple communities.

For eighteen years, I was a member of the Saratoga Rotary, part of Rotary International an organization with 1.4 million members in over 200 countries, dedicated to positively impacting communities globally. Our club of approximately 110 people has given away millions of dollars over the years to different organizations to help people globally. The Rotary motto "Service Above Self" guides humanitarian efforts including polio eradication, literacy programs, and clean water initiatives. It is the intention of International Rotarians that their actions are guided by truth, fairness, and goodwill and are intended to benefit all. I also served as a finance commissioner for the City of Saratoga.

Chapter 22
Personal Growth, Giving Back,
and Reflections

I loved living in the foothills of Saratoga, and buying a home there turned out to be an incredible blessing mentally, physically, spiritually, socially, financially, and professionally. I purchased my Saratoga home from Mr. Bud Beaudoin, and he and I became good friends. Mr. Beaudoin shared stories from his time in public relations at General Electric when actor Ronald Reagan served as GE's television spokesman from 1954 to 1962. He also recalled helping astronaut William Anders, famous for taking the iconic "Earthrise" photograph during the Apollo 8 moon mission in 1968, with a real estate transaction.

Mr. Beaudoin sponsored my membership in the Saratoga Rotary and introduced me to his former son-in-law, a distinguished researcher and professor at MD Anderson Cancer Center, who served as a scientific advisory board member of my company.

To raise funds for cancer patients, I completed the 26.2-mile Napa Valley Marathon at the age of 48 and took part in the American Cancer Society's Relay for Life, a national fundraising event supporting cancer research and patient services. In high school, I led a team of students to walk 20 miles to raise money for the March of Dimes, which has helped millions of premature and sick babies survive and thrive since its founding by President Franklin D. Roosevelt to fight polio. As someone who was born premature and struggled with frequent illnesses as a child, including a near-fatal bout of typhoid fever, completing a marathon held special meaning for me.

While hard work is essential, I believe it's equally vital to enjoy leisure time and live life to the fullest. I've been fortunate to have many adventures over the years, including climbing part of Mount Rainier (14,411 feet), trekking part of the Mount Everest Base Camp trail in Nepal, exploring

Machu Picchu, the ancient Incan city in Peru, ascending Table Mountain in Cape Town, South Africa, trekking Chamonix in the French Alps with a view of Mont Blanc, riding a helicopter to the top of the Mendenhall Glacier in Juneau, Alaska, marveling at Angkor Wat, the world's largest religious monument in Cambodia, and visiting the Great Pyramid of Giza, the last remaining Wonder of the Ancient World in Egypt.

Other highlights included marlin fishing in Cabo San Lucas, exploring Mayan ruins and cenotes in the Yucatán, visiting Egyptian tombs and temples, seeing the Terracotta Army in Xi'an, spending a summer on the French Riviera, relaxing on Hawaiian beaches, and attending the 2008 Summer Olympics in Beijing. It was memorable visiting the Forbidden City, the largest ancient palatial structure in the world at 961 meters by 753 meters, and part of the 13,000-mile Great Wall of China, a UNESCO World Heritage Site.

I enjoyed visiting and learning at world-class museums such as the Louvre in Paris, the British Museum in London, the American Museum of Natural History in New York City, and the Smithsonian Institution Museums in Washington D.C. I especially liked the National Air and Space Museum, where I saw many planes and space vehicles, including the Apollo 11 Command Module "Columbia", which carried astronauts Neil Armstrong, Buzz Aldrin, and Michael Collins to the Moon in July 1969.

Sports and fitness have also been an important part of my life. I played varsity soccer in high school and continued with intramural soccer at UC Berkeley. A highlight was watching the U.S. Women's Soccer Team win Olympic gold in Beijing in 2008. I've enjoyed recreational tennis, cycling, skiing, whale watching, camping in U.S. and Canadian national parks, hiking, and wine tasting. Especially memorable were the Grand Canyon rim-to-Phantom Ranch horseback-and-hiking trip and the trek into the Havasupai Reservation to see the vivid blue pools. Yellowstone, Yosemite, and Denali National Parks were fantastic I could write a chapter on each.

I also enjoyed galloping on the beach in Half Moon Bay and horseback riding on mountain trails. My 100-mile round-trip bicycle ride from Union City through Palo Alto to San Francisco was fun, and the Los Gatos-to-Capitola Beach over the Santa Cruz Mountains ride was a real challenge. I've also enjoyed golf, hang gliding, skydiving, sailing, parasailing, snorkeling, and air shows by the Navy Blue Angels and Air Force Thunderbirds. I once watched a Tour de France stage in Marseille, soaking up the festive crowds.

I'll never forget seeing the America's Cup races on San Francisco Bay or visiting NASA's Ames Research Center, Johnson Space Center, and Kennedy Space Center. Fireworks in Seattle and San Francisco rank among my favorites. Our family ran into Cindy Jung at Expo 86 in Vancouver she's been like a little sister to us for fifty years.

I've witnessed many iconic sports moments: the 49ers winning a Super Bowl, Federer's U.S. Open victory, Tiger Woods at Pebble Beach's U.S. Open, Kobe Bryant's last game at Oracle Arena, the Giants in a private suite, and Usain Bolt's 200 m world record in Beijing 2008.

I'm blessed to have lived and worked in Silicon Valley's epicenter. My career began with a summer job at Siltec in Menlo Park, where on Day 1 I fixed a stubborn wafer-line problem. That early success fueled a lifelong passion for applying engineering to real-world challenges.

Being in the Bay Area meant daily inspiration from Stanford, UCSF, UC Berkeley, and companies like Apple, Google, Meta, Tesla, Intel, Nvidia, HP, Oracle, Salesforce, X, and OpenAI. I've cheered the San Jose Sharks, attended WTA tennis competition at Stanford, and rooted for Cal, Stanford, and Santa Clara in football, basketball, soccer, swimming, and track. Santa Clara's women's soccer titles in 2001 and 2020 were thrilling.

"The Play" in the 1982 Big Game five laterals on the final kickoff with the Stanford Band on the field remains one of college football's most legendary moments.

I've met Steve Young at Santa Clara's Tech Awards and Ronnie Lott at the Cupertino Rotary. Pebble Beach Golf Links, off-tournament, offers unmatched serenity.

When Grace's daughter danced ballet, we attended countless recitals. She couldn't become a Ballerina due to injury, but she graduated summa cum laude in Chemistry from a top private university and got accepted to a top PhD program. I am proud of her.

I love concerts Eagles, Heart, Amy Grant, Celtic Woman, Ringo Starr and once saw Pink Floyd in Marseille. In Prague, street musicians on Charles Bridge gave impromptu orchestra concerts. I also enjoyed seeing Shakira perform at the Latin Grammy.

In Malaysia I bought CDs of a singer whose language I didn't understand proof that music transcends words. I even took a music appreciation class at Berkeley. Taman Negara,

Malaysia's 130-million-year-old rainforest, was another highlight.

Our Santa Clara office sits beside a golf and tennis club, where I play occasionally. From Santa Cruz beaches and Monterey whale watches to Lake Tahoe skiing and skydiving in the Bay Area, I've savored the region's outdoor bounty. I also pilot planes from San Jose International.

Fishing in the Pacific, the Sacramento River, and local lakes rounds out my leisure. Villa Montalvo, eight minutes from home, is a 166-acre park with formal gardens and a hilltop view of Silicon Valley my favorite weekly stroll.

In New York, I've seen the U.S. Open Tennis Finals, the Belmont Stakes (Secretariat's 1973 Triple Crown win by 31 lengths!), the Statue of Liberty, Carnegie Hall, and Central Park.

In 1994 as a student pilot, I survived a low-altitude spin by recalling a training-center video I'd glimpsed an unplanned lesson that saved my life.

Flying over Yosemite Valley and Mono Lake, and piloting to a World Cup match at Stanford Stadium, are cherished memories. There's nothing like the thrill of flight. My dad, my brother and I thoroughly enjoyed the soccer game.

I've owned a Chevrolet Camaro (using my father's GM discount) and a Mercedes-Benz S430 chosen for its blend of performance and comfort, especially for my parents.

When Richard and Grace began college at UC Santa Cruz and UC Davis, I celebrated by gifting each a new car. Richard's charisma and athleticism made him a campus favorite, while Grace's intellect and organization kept our family grounded. Catherine's wit and compassion completed our quartet.

In 1985, Catherine and I bought a Union City home each of our four siblings plus our parents each got a room. When we moved out, we gifted it to Mom and Dad. Forty years later, Zillow valued that house at $1.89 million up from our $201 thousand purchase an illustration of inflation's impact on the dollar.

Chapter 23
A Tribute to My Mother

Of all the remarkable people I've known and loved in my life, none compare to the profound admiration and affection I hold for my mother. Her unconditional love was a constant in my life for 65 years (including the time in the womb) a gift I will forever cherish. She embodied the very best human qualities love, joy, wisdom, kindness, determination, and an indomitable spirit in facing adversity.

In my eyes, she was the epitome of what a perfect mother should be the answer to my every hope and prayer. My mother was her parents' favorite among ten children, a helpful sister, a loving and faithful wife to my father, and a devoted and unselfish mother to my siblings and me. She was well-regarded by her teachers, respected by her students and those she met, and brought joy to the world with her beautiful smile every day. She was also a wonderful cook whose food everyone enjoyed.

My mom was always first in her classes. She was a great 100-meter runner, a talented basketball player, and a wonderful saxophone player who led her own band in high school. She also played violin and piano and sang like a nightingale. Remarkably, she earned a full scholarship to attend National Taiwan University, the most prestigious university in Taiwan, at a time when higher education was rarely accessible to women, especially in STEM fields.

Her passing on July 7, 2023, left a huge void in my heart and an ache that may never fully subside.

My mother's final moments on this earth are forever etched in my memory. As she lay in her bed at home, I sat by her side, holding her hands, my head bowed in silent reverence. Her departure was so peaceful that I didn't even realize she had slipped away until my sister Catherine noticed her breathing had ceased. In that profound moment of grace, mere minutes after her last breath, I witnessed my mother's soul taking flight a luminous essence emanating from the right side of her mouth.

As a man of science, I had always placed faith in empirical evidence and scientific rigor. Yet, in that singular moment, I was granted firsthand proof of the soul's eternal nature a testament to its ongoing journey beyond the physical body. This experience reaffirmed my belief that our time on Earth is but a fleeting instant in the grand tapestry of the soul's infinite voyage, a journey that may continue in perpetuity through the boundless grace and mercy of the Divine.

The COVID-19 pandemic brought an unexpected blessing precious time spent by my mother's side nearly every waking hour, a privilege I will forever cherish. The simple acts of caring for her and sharing daily walks in the park became the centerpiece of my existence. Her radiant spirit and gentle kindness touched everyone she encountered, leaving an indelible mark. I long for her presence intensely, yet take solace in knowing we will reunite in the celestial realms.

It wasn't until grappling with the profound loss of my own mother that I could begin to fathom the depths of grief accompanying such a monumental bereavement. In 2018, when I offered condolences to Dr. Jennifer Doudna upon her mother's passing, I couldn't comprehend the magnitude of the void she must have felt. I remained unaware of Dr. Doudna's 2020 Nobel Prize in Chemistry until after my own mother's passing in 2023, as I was consumed with caring for my mother in her final years.

Chapter 23
A Tribute to My Mother

In summer 2020, my mother was rushed to a COVID-19 intensive care unit after experiencing severe respiratory distress overnight. The attending physicians initially suspected COVID-19, prompting her ICU admission. At the time, hospital policies strictly prohibited family visits to curb the highly infectious disease's spread.

Given the circumstances surrounding my father's recent passing, where we were denied the opportunity to be by his side as he succumbed to COVID-19 in another hospital's ICU, we pleaded for an exception to allow me to see my mother in what we feared were her final moments.

The gravity necessitated I don full personal protective equipment from head to toe before being granted access to her bedside. Around this time, I received a startling call from my Uncle Steven, who informed me that my recently deceased father had appeared to him in a vision, conveying the unsettling news about my mother's hospitalization news I thought only I knew. My father's devotion to my mother was such that he once said he would marry her again in the next life.

Initial COVID-19 tests found my mother negative for the virus. Subsequent tests confirmed this, ruling out coronavirus infection. However, her oxygen levels had plummeted dangerously low to 88%, with fluid accumulating in her lungs. This prompted the medical team to administer a diuretic to alleviate symptoms. Further analysis the next day provided a conclusive diagnosis my mother was suffering from congestive heart failure rather than COVID-19. Thankfully, with proper treatment, she made a full recovery.

Tragically, in early 2020, as the pandemic began sweeping across the United States, my father fell victim to the virus just six days after his diagnosis. At the time, he resided in a prestigious five-star nursing home where COVID-19 ran rampant, infecting every resident and ultimately claiming half their lives. The scourge proved even more pervasive among staff, with more employees infected than residents.

In the pandemic's early stages, the California state government refrained from releasing detailed nursing home data on COVID-19's spread. As a result, I remained unaware of the true extent of the outbreak in my father's care facility.

The nursing home itself implemented strict visitation restrictions and effectively cut off families from loved ones. In a bid to shield these facilities from legal liability, the state government enacted measures protecting them from potential lawsuits over their handling of the pandemic.

According to California Department of Public Health reports, as of mid-2020, nursing homes accounted for a disproportionate share of COVID-19 deaths statewide nearly 50% underscoring systemic vulnerabilities in eldercare facilities nationwide.

Legal protections such as California's Senate Bill 1159 were later enacted to limit liability of healthcare providers during the pandemic, though these raised ethical concerns among families seeking accountability.

Chapter 24
Writing Books and Giving Lectures

At this stage of my life, I am focused on sharing my knowledge to make this world a better place to live. This sharing is urgent and important because the world is facing many unprecedented major challenges simultaneously. We are all fighting against the scourge of pandemics, wars, global financial crises, income inequality, climate change, misuse of artificial intelligence (AI), diseases, unequal opportunities, and overwhelming stress. Many of us are losing our souls. This is our last chance.

The COVID-19 pandemic exposed deep systemic vulnerabilities in public health, global supply chains, and social safety nets, highlighting how interconnected and fragile our modern world truly is.

Meanwhile, ongoing conflicts such as those in Ukraine and the Middle East continue to threaten global stability, while climate change intensifies natural disasters and disrupts ecosystems worldwide.

Economic disparities have widened according to the World Inequality Report 2022, the richest 1% now hold nearly half of global wealth, exacerbating social unrest and limiting access to quality education and healthcare.

Moreover, artificial intelligence, while offering transformative benefits, carries risks when misused or unregulated ranging from privacy violations to deepening societal biases. Responsible development and ethical governance of AI technologies have become critical global imperatives.

I am writing books and giving lectures to make a difference. I hope that you'll join me in this endeavor.

Sections 4: Refocus Your Mind:
A Global Snapshot

Chapter 25
The Origins of the Universe and the Evolution of Life

The universe is a vast and mysterious place, full of wonders that have captivated the human imagination for millennia. From the distant stars and galaxies that light up the night sky to the intricate complexity of life on Earth, the natural world is a source of endless fascination and discovery.

At the heart of our understanding of the universe, most people believe the Big Bang theory, which proposes that the cosmos began as an infinitesimally small, dense, and hot singularity roughly 13.8 billion years ago. In an instant, this primordial point exploded outward, giving rise to all of the matter and energy that we see around us today. As the universe expanded and cooled, subatomic particles began to form and coalesce into the first atoms, primarily hydrogen and helium.

Over millions and billions of years, these atoms were drawn together by the force of gravity, forming vast clouds of gas and dust that eventually gave birth to the first stars and galaxies. The light from these ancient celestial bodies has traveled across the immensity of space and time, reaching us here on Earth and providing a glimpse into the early history of the universe.

Recent observations from the James Webb Space Telescope (JWST) have revealed new insights into the formation and evolution of galaxies in the early universe. The data suggests that some of the earliest galaxies were already forming just 200 million years after the Big Bang, much earlier than previously thought. These primordial galaxies were also more massive and had more complex structures than many scientists had predicted, challenging our understanding of the processes that shaped the cosmos in its infancy.

Dr. John Mather is the Senior Project Scientist for the JWST. Dr. Mather shares the 2006 Nobel Prize for Physics with George F. Smoot of the University of California for their work measuring the heat radiation from the Big Bang. Dr. Mather said that the name Big Bang is really misleading because the universe doesn't have a center, and it happened everywhere all at once, within a process that occurred in time and not at a point in time. He also said that we don't know exactly when the universe made the first stars and galaxies - or how, for that matter. That is what we are building JWST to help answer.

As the universe continued to evolve and expand, the conditions on certain planets became conducive to the emergence of life. Here on Earth, the first living organisms appeared roughly 3.5 billion years ago in the form of simple, single-celled microbes.

Over time, these early life forms may have evolved and diversified, giving rise to the incredible array of species that we see around us today.

On the other hand, some people believe that God made the universe in the beginning. According to the Bible, God said, "Let there be light, and there was light." Suns and stars were thus created.

The process of evolution by natural selection, first described by Charles Darwin in his groundbreaking work "On the Origin of Species," has been a driving force behind the study of biology. According to this theory, individuals within a population that possess traits that are advantageous for survival and reproduction will be more likely to pass on their genes to future generations, leading to the gradual accumulation of beneficial adaptations over time.

Darwin's Theory of Evolution has sparked a great deal of scientific research benefitting mankind and continues to illuminate humanity. For a person who didn't know DNA, his insights were quite remarkable. I appreciate Darwin for his courage and contribution to humanity. However, the theory is incomplete, and we must look at the evidence with clear eyes and have the courage to speak up.

Darwin's work showed that different species of finches evolved from a common ancestor. However, science has not yet been able to explain how non-living matter can give rise to living cells or how complex biological systems can evolve through random mutations alone. DNA, which acts like a complex computer code, guides the development, function, and maintenance of living organisms. The intricate nature of the DNA program suggests it is unlikely to have formed by chance.

Children are taught in school that humans evolved from animals, but there is no conclusive evidence to support this claim. The Gospel of John states, "In the beginning was the Word," which could be interpreted as DNA containing the instructions for life. The same genes are often used in different ways to create a variety of life forms, similar to how a programmer uses small, reusable code snippets to build complex software.

Colossian 1:6 said, "For by him were all things created, that are in heaven, and that are in earth, visible and invisible." No one truly understood what "invisible" meant for nearly two thousand years until Caltech's Fritz Zwicky discovered evidence for dark matter in 1933. In 1998, two teams of astronomers discovered dark energy by measuring light coming from supernovae of type IA. Dark matter and dark energy make up 27% and 68% of the universe, respectively. It is estimated that we are able to see only 5% of the universe.

Genesis 2:7 says, "And the LORD God formed man of the dust of the ground, and breathed into his nostrils the breath of life, and man became a living soul. Darwin's theory of physical evolution from non-life to life, if true, would not result in the creation of an eternal soul. In the universe, physical life is temporary. However, a soul, on the other hand, is eternal and may be capable of traveling anywhere in the universe.

Evolution, as a process, cannot explain the origin of the first living cell. No one believes that the Internet has no creators. A human cell is much more complicated than the Internet. It is not good enough to rely on time and mutations to simply claim that life on Earth has evolved from non-life to life.

Just as people have the right to choose their faith, they should also have the right to choose what makes more sense on the origin of life based on sound reasoning. Science has no convincing answer on dark energy (68%), dark matter (27%), or even how matter (5%) came into existence. The universe was created with a precise expansion rate in an infinitesimally small fraction of a second. Without this fine-tuned expansion, the universe would not exist, as we know it. This rapid expansion was like growing an object of 62 trillion miles in less than a single second. Scientists have confirmed this initial "explosion" through the discovery of cosmic microwave background radiation. Did God create the universe, or did the universe create itself?

The expansion of the universe is accelerating over time, with distant galaxies moving away from us at ever-increasing speeds. The Milky Way, our home galaxy, is immense and ancient. It takes the Sun approximately 250 million years to complete one orbit around the galaxy's center.

The distances between galaxies are much greater than those between stars within a galaxy. Until Edwin Hubble discovered the Andromeda galaxy in 1924, we were unaware of the existence of other galaxies beyond the Milky Way. With the Hubble telescope and JWST, we now know, a mere one hundred years later, that there are more than 200 trillion galaxies in our universe.

Our expanding universe contains many mysteries. It is difficult to imagine that a star thirty times the size of our own sun, which can fit 1.3 million Earth, could collapse into the size of a car in less than 10 seconds in a hypernova and produce enough gamma rays to wipe out almost all life forms on a galactic scale. The Bible describes God creating plants, sea creatures, birds, land animals, and humans on separate (not 24-hour) days. Humans, in particular, are said to have been made in God's image.

Some scientists suggest that human chromosome 2 resulted from the fusion of two primate chromosomes. However, creationists argue that there is no fossil evidence to support the idea of humans evolving from a common ancestor with apes. Science has not yet explained what triggered the Big Bang or the origins of matter, energy, and space.

People's beliefs about the origin of life often depend on their initial assumptions. Darwin's theory does not address how the first cell or DNA came into existence. Each organism develops according to its unique DNA instructions, and science has not demonstrated how a functional cell or complex DNA could arise by chance. The existence of a living cell is a prerequisite for evolution to occur.

Human life begins when the sperm of a man and the egg of a woman are combined to form what could become a newly conscious and soulful individual. Each of us begins as a single cell and a single word with three billion base pairs that already contain the blueprint of who we will become in due time. Human sperm and egg must meet specific requirements that a random process is unlikely to produce. If your life depends on it, and it does, would you depend on a random process or God to make you into a human?

A unique word, a series of chemical molecules, and a life program called DNA distinguish each individual from the other. As we begin our lives on a blue planet among trillions of planets in our galaxy, all of these aspects of our nature have already been determined at the moment of conception, from the color of our eyes to the temperament of our personality, including our inherited susceptibility to various diseases.

The majority of our bodies are composed of salt water. Liquids expand on heating as their molecules move with greater energy, overcoming intermolecular attraction. Liquids usually contract on cooling as the molecules move slower and are unable to overcome the force of attraction between them. When they freeze, they contract more to form rigid solid structures with minimal intermolecular spaces. In contrast, water expands by 9-10% when it freezes, which is why ice floats on water instead of contracting. If ice sinks, life would be greatly impacted. Evidence for a global flood can be found in flood stories from cultures worldwide and the presence of seashells on mountaintops.

According to the Bible, Noah's ark carried Noah, his family, and representatives of various land animals. Since Noah lived less than 10,000 years ago, I don't think there were dinosaurs in the ark. After the flood, human lifespans were limited to 120 years. Today, it is rare to find individuals living beyond 120 years.

Research in biology has shown that when all of the cells ever created in the human body are multiplied by the average time it takes for cells to reach the end of their lives, you get roughly 120 years. This is called the "Hayflick limit, the maximum number of years a human can expect to live." Leonard Hayflick is a professor of anatomy at the University of California, San Francisco, and formerly a professor of medical microbiology at Stanford University.

Elizabeth H. Blackburn at the University of California in San Francisco and Jack W. Szostak at Harvard Medical School in Boston, Massachusetts, also applied Hayflick's theory of cellular aging to their research on the structures of telomeres in 1982 when they cloned and isolated telomeres. In 2009, Blackburn and Szostak received the Nobel Prize in Physiology or Medicine for their work on telomerase, in which the Hayflick Limit played an essential role.

While science and religion can sometimes seem to be in conflict, some scientists and theologians argue that the two domains of knowledge are ultimately compatible. Science seeks to understand the natural world through observation, experimentation, and reason, while religion deals with questions of meaning, purpose, and the nature of the divine. Both can offer valuable insights into the human experience and the mysteries of the universe, and both have played important roles in shaping the course of human history.

As we continue to explore the origins and evolution of the universe and life on Earth, it is clear that there is still much that we do not understand. From the nature of dark matter and dark energy to the question of whether we are alone in the universe, the frontiers of science are filled with profound mysteries and unanswered questions. But perhaps the greatest mystery of all is the nature of consciousness and the human mind. How is it that a collection of atoms and molecules, arranged in just the right way, can give rise to the subjective experience of awareness, thought, and emotion? Is consciousness purely a product of the brain, or is there some immaterial aspect of the mind that transcends the physical world?

These are questions that have puzzled philosophers and scientists for centuries, and they continue to inspire intense debate and investigation today. Some researchers believe that consciousness can be fully explained in terms of the complex interactions of neurons in the brain, while others argue that there must be some additional factor at work, such as a non-physical soul or spirit.

Regardless of one's personal beliefs about the nature of consciousness, it is clear that the human mind is one of the most remarkable and mysterious aspects of the universe. Through the power of thought and imagination, we have been able to unravel the secrets of the cosmos, create works of art and literature that move the soul, and build civilizations that span the globe.

As we stand at the threshold of a new era of scientific discovery and exploration, it is more important than ever that we approach the great questions of existence with humility, curiosity, and an open mind. We must be willing to follow the evidence wherever it leads, even if it challenges our preconceived notions or deeply held beliefs. At the same time, we must also recognize the limits of our understanding and the vastness of the mysteries that still remain.

In the end, the story of the universe and the evolution of life on Earth is a tale of incredible beauty, complexity, and wonder. From the first moments of the Big Bang to the emergence of human consciousness, forces and processes that are both awe-inspiring and humbling in their scale and power have shaped the natural world. As we continue to explore the frontiers of science and push the boundaries of human knowledge, let us never lose sight of the incredible privilege and responsibility we have as conscious beings in a vast and ancient cosmos. Let us approach the great questions of existence with a sense of reverence and wonder, and let us work to build a future that is worthy of the incredible legacy of the universe that birthed us.

If we took one second to name each sun in the universe, it would take at least ten thousand trillion years. We live in a very, very big universe. It is improbable to think that God created life only on Earth. The universe will go on without humans. Earth will go on without humans, but humans cannot experience the fullness of life without Earth or something like Earth.

Chapter 26
The Rise of Civilization and the Origins of Religion

The emergence of human civilization is a story of ingenuity, perseverance, and the remarkable ability of our species to adapt and thrive in a wide range of environments. From the earliest hunter-gatherer societies to the great empires of the ancient world, the history of human culture is a testament to the power of the human mind and spirit.

The origins of human civilization can be traced back to the end of the last ice age, roughly 12,000 years ago. As the climate began to warm and the glaciers retreated, our ancestors found themselves in a world that was ripe for exploration and settlement. The development of agriculture and the domestication of animals allowed for the rise of permanent settlements and the growth of complex societies.

One of the earliest and most influential civilizations to emerge was that of ancient Mesopotamia, located in the Fertile Crescent between the Tigris and Euphrates rivers. The Sumerians, who lived in this region from roughly 4500 to 1900 BCE, are often credited with the invention of writing, the wheel, and the first system of laws. The earliest known writing system, cuneiform, emerged around 3200 BCE in Sumer and was initially used for accounting and administrative purposes. They also had a complex pantheon of gods and goddesses, each associated with different aspects of the natural world and human society.

The ancient Egyptians, who built their civilization along the banks of the Nile River, are another example of an early society that made significant contributions to human culture. The Egyptians are known for their impressive feats of engineering, including the construction of the pyramids and the development of a sophisticated system of irrigation and agriculture. They also had a rich tradition of art, literature, and religion, with a pantheon of gods and goddesses that included Ra, the sun god, and Osiris, the god of the underworld. Egyptian hieroglyphics, developed around 3100 BCE, represent one of the oldest writing systems and were used extensively in religious texts and inscriptions.

Sumerian and Egyptian writing systems are among the oldest known, dating back to approximately 3200 BCE. Chinese writing, which developed around 1500 BCE, is likely the oldest continuously used writing system in the world.

The Indus Valley Civilization, also known as the Harappan Civilization, thrived in present-day Pakistan and India between approximately 3300 and 1300 BCE, with its mature phase lasting from 2600 to 1900 BCE. While it is older than the classical periods of Mesopotamian and Egyptian civilizations, less is known about its origins, development, and decline.

The discovery of the Indus Valley Civilization's large, well-planned cities in the 1800s revealed that many of the "firsts" attributed to Mesopotamia and Egypt may actually belong to this civilization. Sites like Mohenjo-daro and Harappa featured advanced urban planning, drainage systems, and standardized weights and measures suggesting a high degree of civic organization.

Remains of sophisticated pre-agricultural structures dating back to around 9600 BCE have been discovered at sites such as Göbekli Tepe in modern-day Türkiye. These megalithic temples, intentionally buried and only excavated in the late 20th century, challenge conventional timelines of the rise of civilization and suggest that complex societal structures may have existed prior to the development of farming.

In ancient Greece, religion and mythology played a central role in society. While Greek gods are often seen today as mythological explanations for natural phenomena, there is evidence both for and against their existence.

Most people would consider the Trojan War and the Trojan Horse to be purely mythological if the city of Troy had not been discovered. The ruins of Troy, located in modern-day Hisarlik, Türkiye, were identified by Heinrich Schliemann in the 1870s, providing archaeological evidence that the legendary conflict may have been rooted in historical events dating back to the 12th or 13th century BCE.

The oldest known written history comes from Sumer in Mesopotamia, recorded on clay tablets. These tablets describe a king named Gilgamesh, who is believed to have ruled around 2700 BCE. The stories of Gilgamesh and his epic adventures have become legendary. The "Epic of Gilgamesh" is widely regarded as the earliest surviving great work of literature, offering profound insights into ancient views on mortality, kingship, and human nature.

As human societies continued to grow and develop, they began to interact with one another through trade, migration, and conquest. The rise of empires such as those of the Persians, Greeks, and Romans brought people from different cultures and backgrounds into contact with one another, leading to the exchange of ideas, technologies, and ways of life.

One of the most significant developments in the history of human civilization was the emergence of the first great monotheistic religions, such as Judaism, Christianity, and Islam. These faiths, which share a common belief in a single, all-powerful God, have had a profound impact on the course of human history and continue to shape the lives of billions of people around the world today.

The origins of Judaism can be traced back to the ancient Israelites, a Semitic people who lived in the region of Canaan (modern-day Israel and Palestine) in the second millennium BCE. According to Jewish tradition, the Israelites were chosen by God to be his special people and were given the Torah, a set of sacred laws and teachings that formed the basis of their religion and way of life. Archaeological and textual evidence suggests that the Kingdoms of Israel and Judah emerged by the 10th century BCE, and the Torah was compiled over several centuries, with core texts finalized during the Babylonian exile (6th century BCE).

Christianity, which emerged in the first century CE as a sect within Judaism, is based on the life and teachings of Jesus Christ, whom Christians believe to be the Son of God and the savior of humanity. The spread of Christianity throughout the Roman Empire and beyond had a profound impact on the development of Western civilization, shaping everything from art and literature to politics and social norms. By the 4th century CE, Christianity became the dominant religion of the Roman Empire under Emperor Constantine, leading to the establishment of the Church as a major institution in European history.

Islam, which emerged in the seventh century CE in the Arabian Peninsula, is based on the teachings of the Prophet Muhammad and the belief in a single, all-powerful God

(Allah in Arabic). Like Judaism and Christianity, Islam has played a significant role in shaping the course of human history, particularly in the Middle East, North Africa, and parts of Asia and Europe. The Islamic Golden Age (8th to 13th centuries) witnessed major advances in science, medicine, mathematics, and philosophy, with centers of learning such as Baghdad's House of Wisdom influencing both Eastern and Western intellectual traditions.

While these three monotheistic faiths share many common beliefs and values, they have also been the source of significant conflict and division throughout history. The Crusades, which were a series of religious wars fought between Christians and Muslims in the Middle Ages, are one example of how religious differences can lead to violence and bloodshed. Beginning in 1096 CE and lasting for nearly two centuries, the Crusades were initially launched by the Catholic Church to reclaim Jerusalem and other holy sites from Muslim control. These campaigns resulted in widespread violence, not only between Christians and Muslims but also against Jewish communities in Europe, highlighting the complexities and consequences of religious zeal and political ambition.

In addition to these major world religions, human societies have also developed a wide range of other spiritual and philosophical traditions over the course of history. From the animistic beliefs of indigenous peoples to the meditation practices of Buddhism and Hinduism, the human search for meaning and purpose has taken many different forms across cultures and time periods.

Buddhism, founded in the 5th to 6th century BCE by Siddhartha Gautama in India, emphasizes mindfulness, ethical conduct, and the cessation of suffering. Hinduism, one of the world's oldest religions, has roots that trace back over 4,000 years and encompasses a vast array of deities, texts, and philosophies including the Vedas, Upanishads, and the Bhagavad Gita. Indigenous spiritual systems such as Native American, Aboriginal Australian, and African tribal religions often emphasize a deep connection with nature, ancestor reverence, and cyclical time.

One of the most enduring questions in the study of religion is the relationship between faith and reason. While some religious traditions emphasize the importance of blind faith and obedience to divine authority, others encourage the use of reason and critical thinking in the pursuit of spiritual truth. For example, medieval Islamic philosophers such as Avicenna (Ibn Sina) and Averroes (Ibn Rushd) worked to harmonize Greek philosophy with Islamic theology, influencing both the Islamic Golden Age and European scholasticism. In Christianity, figures like Thomas Aquinas attempted to reconcile Aristotelian logic with Christian doctrine in his seminal work Summa Theologica.

The Scientific Revolution of the 16th and 17th centuries, which led to the development of modern science, posed a significant challenge to traditional religious beliefs and authority. As scientists began to uncover the laws and mechanisms that govern the natural world, some religious leaders saw this as a threat to their power and influence.

Key figures of the revolution included Copernicus, Kepler, Newton, and Galileo who challenged geocentric cosmology, expanded astronomical understanding, and laid the foundations for classical physics. Their findings often conflicted with Church teachings that interpreted the Bible literally, leading to doctrinal disputes and institutional pushback.

However, many scientists and philosophers have argued that science and religion are not necessarily in conflict with one another but rather address different aspects of the human experience.

This view is often referred to as the "non-overlapping magisteria" model, proposed by evolutionary biologist Stephen Jay Gould, which suggests that science and religion occupy distinct domains of teaching authority.

One of the most famous examples of this perspective is the work of Galileo Galilei, the Italian astronomer and physicist who is often credited with the invention of the telescope. Despite facing significant opposition from the Catholic Church for his support of the Copernican model of the solar system, Galileo argued that science and religion could coexist peacefully as long as each stayed within its proper domain.

Galileo did not invent the telescope but significantly improved it and used it to support heliocentric theory. In 1633, he was tried by the Roman Inquisition and placed under house arrest, yet he continued to write. His famous assertion that "the Bible teaches us how to go to heaven, not how the heavens go" captures his view of the distinction between theological and scientific truths.

Today, the relationship between science and religion remains a complex and often contentious issue, with some people seeing them as fundamentally incompatible and others arguing for a more harmonious integration of the two. Regardless of one's personal beliefs, it is clear that both science and religion have played a significant role in shaping human civilization and culture throughout history.

Modern thinkers such as Albert Einstein, Carl Sagan, and Teilhard de Chardin have contributed to the ongoing dialogue, exploring ways that scientific wonder and spiritual insight can complement one another. Einstein, for instance, famously remarked, "Science without religion is lame, religion without science is blind," illustrating the nuanced interplay between empirical knowledge and metaphysical belief.

As we face the many challenges and opportunities of the 21st century and beyond, let us draw on the wisdom and insights of the great spiritual and philosophical traditions that have come before us while also remaining open to new ideas and discoveries that may challenge or transform our understanding of the world and our place within it.

In an age marked by climate change, artificial intelligence, and genetic engineering, the intersection of ethical reflection and empirical inquiry will become increasingly vital. A balanced embrace of both ancient wisdom and scientific progress may help humanity navigate the moral and existential dilemmas of the future.

Chapter 27
Natural World's Marvels and
Universe's Mysteries

From the tiniest subatomic particles to the vast expanse of the cosmos, the natural world is a source of endless wonder and mystery. For centuries, humans have sought to understand the workings of the universe, to uncover the secrets of life and matter, and to marvel at the beauty and complexity of the world around us.

At the heart of our understanding of the natural world is the science of physics, which seeks to explain the fundamental laws and forces that govern the behavior of matter and energy. From Newton's laws of motion to Einstein's theory of relativity and Bohr's quantum theory, the insights of physics have revolutionized our understanding of the universe and our place within it. These theories laid the foundation for modern science and led to advancements in space exploration, nuclear energy, and modern electronics.

One of the most remarkable discoveries of modern physics is the strange and counterintuitive world of quantum mechanics. At the subatomic level, particles can exist in multiple states at once, and the act of observation can actually influence the outcome of an experiment.

Scientists are still unraveling the implications of quantum mechanics today, but the efforts have already led to the development of technologies such as lasers, transistors, and magnetic resonance imaging (MRI). Additionally, quantum theory is central to ongoing innovations in quantum computing, which promises to exponentially increase processing power beyond classical limits.

Another area of physics that has captured the imagination of scientists and the public alike is the study of the universe as a whole. From the Big Bang theory to the discovery of dark matter and dark energy, cosmologists have made incredible progress in understanding the origins and evolution of the cosmos.

The Big Bang, estimated to have occurred 13.8 billion years ago, marked the beginning of space, time, and matter. However, around 95% of the universe remains unexplained, made up of dark matter (27%) and dark energy (68%) components inferred from gravitational effects and cosmic expansion.

One of the most exciting developments in the field of astronomy in recent years has been the discovery of exoplanets planets that orbit stars other than our own sun. Thanks to powerful new telescopes and detection methods, astronomers have now confirmed the existence of thousands of these worlds, ranging from rocky, Earth-like planets to massive gas giants that dwarf even Jupiter in size.

As of 2025, over 5,600 confirmed exoplanets have been cataloged in more than 4,000 planetary systems, with data from missions like Kepler, TESS, and the James Webb Space Telescope contributing significantly to this discovery.

The study of exoplanets has opened up new possibilities for the search for life beyond Earth. While we have yet to find definitive evidence of extraterrestrial life, the sheer number and diversity of exoplanets suggest that the ingredients for life may be common throughout the universe.

Particularly promising are planets located in the "habitable zone" regions where temperatures allow for the presence of liquid water, a key ingredient for life as we know it.

But the wonders of the natural world are not limited to the far reaches of space. Here on Earth, the study of biology has revealed the incredible complexity and diversity of life in all its forms. From the intricate workings of the human body to the delicate balance of ecosystems, the science of life is a testament to the power of biology and the resilience of nature.

Recent advances such as CRISPR gene-editing, microbiome research, and synthetic biology have expanded our ability to understand, modify, and protect living systems offering new solutions for health, agriculture, and environmental sustainability.

One of the most remarkable examples of the complexity of life is the human brain. With its trillions of neural connections and its ability to process vast amounts of information, the brain is perhaps the most sophisticated and mysterious organ in the known universe.

The study of neuroscience has made incredible progress in understanding the workings of the brain, from the molecular level to the level of consciousness and cognition.

Recent advances in brain imaging techniques, such as fMRI and PET scans, have allowed researchers to observe neural activity in real time, linking brain regions to behavior, memory, emotion, and decision-making.

Another area of biology that has captured the imagination of scientists and the public alike is the study of genetics.

The discovery of the structure of DNA by Watson and Crick in 1953 revolutionized our understanding of the basic building blocks of life and paved the way for incredible advances in fields such as medicine, agriculture, and forensic science. Their model, a double helix structure, was built on data gathered by Rosalind Franklin through X-ray crystallography an essential but often undercredited contribution.

Today, scientists are using the tools of genetics to unravel the secrets of human history, develop new treatments for diseases, and create new forms of life in the lab. The possibilities of genetic engineering are both exciting and controversial, raising profound questions about the nature of life and the ethical implications of our growing power to manipulate it.

The Human Genome Project, completed in 2003, decoded the entire human genetic blueprint, and new gene-editing technologies like CRISPR-Cas9 now allow precise alterations to DNA, offering hope for curing genetic disorders but also sparking debates on designer babies, bioethics, and unintended ecological consequences.

But for all the progress that science has made in understanding the natural world, there is still much that remains unknown and unexplored. While evolution through natural selection is the widely accepted mechanism by which life developed complexity, the DNA program's intricate coding and regulatory systems have led some to speculate on deeper questions of origin.

From the mysteries of dark matter and dark energy to the question of the origin of life itself, the frontiers of science are filled with unanswered questions and uncharted territories.

For example, abiogenesis the theory of how life arose from non-living matter remains an area of intense study, with researchers exploring prebiotic chemistry and RNA-world hypotheses to bridge the gap between chemistry and biology.

These are questions that have puzzled philosophers and scientists for centuries, and they continue to inspire new avenues of research and exploration today. Some scientists believe that the key to understanding consciousness may lie in the strange world of quantum mechanics, while others look to the study of artificial intelligence and machine learning for insights into the nature of the mind.

Though controversial, theories such as Orch-OR (orchestrated objective reduction), proposed by Roger Penrose and Stuart Hameroff, explore the possibility of quantum processes in brain microtubules as contributors to consciousness. Meanwhile, AI models inspired by neural networks are helping to simulate and study decision-making and perception.

Regardless of the specific approach, it is clear that the study of consciousness and the human experience is one of the great scientific and philosophical challenges of our time. It is a challenge that requires not only rigorous research and experimentation but also a deep appreciation for the mysteries and wonders of the natural world. Interdisciplinary efforts in neuroscience, psychology, physics, and philosophy continue to push the boundaries of what we understand about self-awareness and subjective experience.

In the end, the beauty and complexity of the universe are not something that can be fully captured by equations or experiments alone. It is something that must be experienced firsthand through the lens of human perception and imagination.

Whether we are gazing up at the stars in wonder or marveling at the intricacies of a single living cell, the natural world has the power to inspire awe, curiosity, and a deep sense of connection to something greater than ourselves. This sense of wonder has historically fueled scientific revolutions from Galileo's telescope to Darwin's voyage on the Beagle and continues to shape our collective human story.

As we continue to explore the frontiers of science and push the boundaries of human knowledge, let us never lose sight of the incredible privilege and responsibility we have as conscious beings in a vast and mysterious universe. Let us approach the great questions of existence with humility, passion, and an open mind, and let us work to build a future that is worthy of the incredible legacy of the natural world that sustains us.

In the end, the study of the marvels of the natural world and the mysteries of the universe is not just a scientific endeavor, but also a deeply human one. It is a pursuit that has the power to unite us across cultures and generations, to inspire us to dream big, to imagine new possibilities, reminding us of the fragility and incredible beauty of the world we share.

Global scientific initiatives like climate research, space exploration, and pandemic response demonstrate the unifying power of collaborative knowledge in addressing shared challenges.

Chapter 27
Natural World's Marvels and
Universe's Mysteries

So let us embrace the wonders and mysteries of the universe with all our hearts and minds, and let us work together to build a future that is rich in knowledge, compassion, and stewardship of the natural world. In the end, it is only by understanding and cherishing the world around us that we can hope to create a brighter and more sustainable future for generations to come. This includes prioritizing sustainability in energy use, biodiversity conservation, and ethical innovation in science and technology.

In today's world, we have access to an unprecedented wealth of information through books, recordings, and digital media. As human beings, we possess a sense of morality and conscience that sets us apart from other creatures. We strive to understand the world around us and to live lives filled with meaning and purpose. Anthropological studies suggest that symbolic thought, language, and cooperative social behavior traits deeply tied to moral awareness evolved alongside the development of the human brain and cultural complexity.

Our ability to walk upright and use our hands with great dexterity has allowed us to create tools and build complex structures. We cultivate crops and raise animals for food, and we have developed clothing to provide warmth, protection, and a means of self-expression.

The control of fire, invention of the wheel, and development of agriculture during the Neolithic Revolution marked major milestones in early human civilization, ultimately leading to cities, art, governance, and technological innovation.

Dr. James Watson, one of the co-discoverers of the DNA double helix structure, does not believe in the existence of God. He has been quoted as saying, "If we don't play God, who will?" This statement reflects his belief in the power of science to shape human destiny and his skepticism toward supernatural explanations.

While Dr. Watson and his colleague Dr. Francis Crick received the Nobel Prize in Physiology or Medicine in 1962 for their groundbreaking work on DNA, it is important to note that their discovery built upon the crucial, not-yet-published research conducted by Dr. Rosalind Franklin. Her X-ray diffraction images, especially the famous "Photo 51," provided critical evidence that helped confirm the helical structure of DNA.

Sadly, Dr. Franklin died of ovarian cancer in 1958 at the age of 37, before she could be fully recognized for her contributions. Because the Nobel Prize is not awarded posthumously, her role was formally overlooked, though modern scholarship has since brought her contributions to light.

The fact that even the most brilliant scientists can disagree about the existence of a higher power highlights the ongoing debate surrounding the role of faith in the modern world. For evolutionary processes to occur, species must adapt to their environments, accumulate genetic mutations, and undergo epigenetic changes that alter gene expression patterns.

Epigenetics, which refers to chemical modifications that influence gene activity without changing the underlying DNA sequence, plays a key role in development and adaptation. For many people, belief in God is as fundamental to their worldview as concepts like love and hope. This coexistence of scientific inquiry and spiritual belief reflects the diverse ways in which humans seek to understand meaning, purpose, and the nature of existence.

Chapter 28
Search for Purpose and the Future of Our Species

As we stand at the threshold of a new era in human history, it is impossible not to feel a sense of both excitement and trepidation about what the future may hold. On the one hand, the incredible advances of science and technology have opened up new frontiers of knowledge and possibility, offering the promise of a world that is healthier, more prosperous, and more connected than ever before.

From CRISPR gene-editing to artificial intelligence, quantum computing, and renewable energy breakthroughs, our capacity to transform life on Earth is accelerating at an unprecedented pace. On the other hand, the challenges facing our species from the pandemic, climate change, and environmental destruction to social inequality and the threat of nuclear war have never been more urgent or complex.

At the heart of these challenges lies a deeper question about the nature and purpose of human existence itself. For millennia, philosophers, theologians, and scientists have grappled with the question of what it means to be human and what our role and responsibility are in the grand scheme of the universe.

Some have argued that the purpose of human life is to seek knowledge and understanding, to unravel the mysteries of the natural world, and to use that knowledge to better the human condition. This drive has led to monumental achievements from decoding the human genome to launching the James Webb Space Telescope, which now peers back over 13 billion years to the early universe.

Chapter 28
Search for Purpose and the
Future of Our Species

Others have seen the essence of humanity in our capacity for love, compassion, and creativity and have argued that our highest calling is to cultivate these virtues and work towards a more just and harmonious world.

Still, others have found meaning and purpose in the pursuit of spiritual or religious truth, in the belief that there is a higher power or divine plan that gives shape and direction to our lives. And for many, the search for purpose is a deeply personal and individual journey, one that is shaped by our unique experiences, beliefs, and values.

According to recent global surveys, such as the Pew Research Center's Religion & Public Life study, the majority of the world's population continues to adhere to some form of religious or spiritual worldview, suggesting that the quest for higher meaning remains a universal human trait.

Regardless of our specific beliefs or philosophies, it is clear that the question of human purpose and meaning is one that is central to our existence as a species. It is a question that has driven the great achievements and struggles of human history, from the rise of great civilizations to the fight for social justice and human rights.

Movements such as the Enlightenment, civil rights campaigns, and the global push for sustainable development all reflect this ongoing human effort to align knowledge with moral and social progress.

As we look to the future, it is a question that will continue to shape the course of human events in profound and unpredictable ways. With the rapid pace of technological change and the growing challenges of the 21st century, the stakes have never been higher for our species.

One of the most pressing issues facing humanity today is the threat of climate change and environmental destruction. The scientific evidence is clear that human activities from the burning of fossil fuels to deforestation and habitat destruction are driving rapid and unprecedented warming of the planet, with potentially catastrophic consequences for both human society and the natural world.

According to the Intergovernmental Panel on Climate Change (IPCC), global temperatures have already risen approximately 1.1°C above pre-industrial levels, and current trajectories could lead to increases of 2°C or more by the end of the century if emissions are not drastically reduced. This rise is driving more frequent and intense extreme weather events, contributing to sea-level rise, causing biodiversity loss, and creating challenges for food and water security.

To address this crisis, we will need to fundamentally rethink the way we live and work, transitioning away from an economy based on fossil fuels and unsustainable consumption towards one that is built on renewable energy, sustainable agriculture, and circular production systems. The International Energy Agency (IEA) estimates that to reach net-zero emissions by 2050, global renewable energy capacity must triple, alongside widespread electrification and energy efficiency improvements. This will require not only technological innovation and policy change but also a profound shift in values and priorities towards a new ethic of environmental stewardship and responsibility.

At the same time, we will need to confront the deep social and economic inequalities that persist both within and between nations and that threaten to undermine the stability and cohesion of our global society. From the widening gap between rich and poor to the ongoing struggles for racial and gender equality, the challenges of building a more just and equitable world are as urgent and pressing as ever. According to the World Inequality Report 2022, the richest 10% of the global population earn approximately 52% of global income, while the poorest 50% earn just 8.5%. Such disparities contribute directly to social unrest and reduced opportunities for millions.

To meet these challenges, we will need to embrace a new vision of human purpose and potential one that recognizes the inherent dignity and worth of every individual and that seeks to create a world in which everyone has the opportunity to thrive and reach their full potential. This will require a renewed commitment to education, healthcare, and social welfare, as well as a willingness to challenge entrenched systems of power and privilege. The United Nations' Sustainable Development Goals (SDGs) provide a framework to address these issues, aiming to eradicate poverty, achieve quality education for all, and promote gender equality by 2030.

However, perhaps the greatest challenge facing humanity in the coming centuries is the question of our relationship with technology and the implications of rapid advances in fields such as artificial intelligence, biotechnology, and robotics. As these technologies continue to evolve and become more sophisticated, they hold both incredible promise and potential peril for our species.

On the one hand, advances in AI and robotics could help us to solve some of the greatest challenges facing humanity, from curing disease and extending human life to exploring the frontiers of space and unlocking the secrets of the universe. For example, AI-driven drug discovery platforms have accelerated vaccine development processes, and robotic technologies are transforming surgery with precision beyond human capability.

On the other hand, the development of superintelligent AI machines that surpass human intelligence could pose existential risks to our species, raising profound questions about the nature of consciousness, free will, and the future of humanity itself.

Experts like Nick Bostrom and organizations such as OpenAI emphasize the importance of ethical AI development and global cooperation to mitigate risks associated with advanced AI systems.

As we grapple with these challenges and opportunities, it is clear that the future of our species will depend on our ability to adapt and evolve, both technologically and socially. We will need to find new ways to live and work together, build resilient and sustainable communities, and harness the power of science and technology for the greater good.

Above all, we will need to cultivate a deeper sense of purpose and meaning in our lives, one that goes beyond the pursuit of material wealth or individual achievement and that recognizes the profound interconnectedness of all life on Earth. Whether through the practice of mindfulness and compassion, the pursuit of knowledge and understanding, or the cultivation of creativity and wonder, we must each find our own path to a life of purpose and fulfillment.

In the end, the search for purpose and meaning is a deeply human endeavor, one that has shaped the course of our species from the very beginning. And as we stand at the threshold of a new era of human history, it is an endeavor

Chapter 28
Search for Purpose and the
Future of Our Species

that will continue to define us, both as individuals and as a global community.

Chapter 29
The Bible and Its Interpretations

The Bible contains a passage that describes the "sons of God" marrying human women and giving birth to legendary heroes and warriors, often referred to as the Nephilim (Genesis 6:1-4). There is much debate among scholars about the identity of these "sons of God," with some arguing that they were fallen angels or divine beings, a view supported in ancient Jewish texts like the Book of Enoch, which is considered apocryphal by most Christian denominations. Others maintain that they were simply righteous men or descendants of Seth. This debate is complicated further because other biblical verses (e.g., Romans 8:14) refer to believers as "sons of God," indicating a metaphorical use of the term.

According to the Bible, God formed Adam from the dust of the earth and breathed life into him, making him a living being with a soul (Genesis 2:7). This account differs from Darwin's theory of evolution, which focuses solely on physical changes and natural selection and does not address the development of consciousness or the soul. Many theologians argue that the soul is a spiritual entity, eternal in nature, capable of existing independently of the physical body, transcending the limitations of space and time, a concept supported by various religious traditions but not empirically verifiable by science.

Biblical scholars like Bishop James Ussher have attempted to calculate the age of the Earth based on a literal interpretation of the Bible. Ussher famously placed the date of creation at 4004 BCE and the occurrence of Noah's flood at 2349 BCE. However, other chronologies, such as those based on the Septuagint or Samaritan Pentateuch, suggest different dates, with some placing the flood around 2459 BCE. Modern geology and radiometric dating place the Earth's age at approximately 4.54 billion years, highlighting the tension between literal biblical chronologies and scientific consensus.

Despite the fact that Christianity has its roots in Judaism and Jesus himself was Jewish, there are some Christians who express hostility towards Jews and the Jewish faith. This unfortunate reality has led to historical tensions, including antisemitism, which has been widely condemned by modern Christian denominations. Jesus emphasized that his mission was to fulfill the teachings of the Old Testament, not to abolish them (Matthew 5:17). However, many Jews do not accept Jesus as the Messiah, a fundamental theological divergence between Judaism and Christianity.

Israel's survival in the face of threats from neighboring countries is often attributed to the support it receives from Christians who believe in the biblical importance of the Jewish people, a phenomenon sometimes called Christian Zionism. Many Arabs, on the other hand, view the modern state of Israel as a foreign presence in a region that has been predominantly Arab for centuries. They see Israel as a problem that needs to be resolved, contributing to ongoing geopolitical conflicts that date back to the mid-20th century and earlier.

Throughout history, there have been those who dismiss end-times prophecies as mere fantasy, while others live in constant anticipation of the world's end. Jesus himself stated that no one knows the exact time of his return except for God the Father (Mark 13:32). As a result, numerous attempts to predict the precise date of the end times have proven to be misguided and false. In an age of information overload and conflicting ideologies, discerning truth from falsehood can be a daunting task, especially regarding apocalyptic teachings.

No one can escape physical death, not even Jesus, who Christians believe to be the incarnation of God (Luke 23:46). While many people aspire to grow in their understanding of God's love, intellectual knowledge alone is insufficient. Developing a deep, personal relationship with God requires active commitment and a willingness to live out one's faith. This is emphasized in passages such as James 2:17, which states that faith without works is dead.

It is said that even Satan, who was once an angel of light (2 Corinthians 11:14), possesses a greater understanding of God's nature than any human being. However, knowledge without righteousness is ultimately meaningless. To truly follow God, one must not only recognize the difference between good and evil but also have the courage to act on that understanding.

As artificial intelligence continues to advance, it is possible that those who develop these technologies will wield enormous power over humanity, raising new ethical and spiritual questions about stewardship and responsibility.

As someone who has spent years studying the Bible, I believe it to be the most influential and transformative book in human history. However, I also recognize that the Bible contains passages that can easily be misinterpreted or taken out of context, leading to confusion and error.

For example, cultural, historical, and linguistic contexts are crucial for accurate interpretation, and neglecting these can result in significant misunderstandings. In the following pages, I will present evidence to support my perspective on these matters.

The concept of a divine trinity, or three persons in one God, is not unique to Christianity. Buddhism has the trinity of Amitabha Buddha and his two bodhisattvas, while Hinduism has the trinity of Brahma, Vishnu, and Shiva. In Christianity, the trinity consists of God the Father, God the Son (Jesus), and God the Holy Spirit. These three distinct entities are understood to be united in one divine being (Matthew 28:19).

The doctrine of the Trinity was formally articulated in the Nicene Creed (325 AD) as a response to various early Christological debates. God is everywhere at the same time (omnipresence). To a lesser extent, the Internet exists almost everywhere around the world at the same time.

The Old Testament portrays God as the creator of the universe, who formed the world in six days and later sent a great flood to purge the earth of wickedness (Genesis 1–7). Some scholars argue that the term "day" (Hebrew yom) in this context does not necessarily refer to a 24-hour period but could represent a much longer span of time, a view known as the Day-Age Theory, which attempts to reconcile the biblical account with scientific evidence for an ancient Earth. On Day 1 of creation, Earth was still without form and void (Genesis 1:2).

The God of the Old Testament is often depicted as a stern and demanding deity who expects strict obedience from his followers. He is referred to by the name Yahweh (often represented as YHWH) and is associated with laws such as "an eye for an eye" (Exodus 21:24) and dietary restrictions found in Leviticus.

Young Earth creationists take the Genesis account of creation literally, believing that the universe is only about 6,000 years old. They argue that scientific evidence pointing to a much older universe is being misinterpreted and that radiometric dating methods are fundamentally flawed. They also claim that the existence of soft tissue in dinosaur fossils is proof that these creatures lived relatively recently rather than at least 65 million years ago. This claim remains highly controversial and widely disputed among paleontologists, with many experts attributing soft tissue preservation to rare but natural chemical processes.

In contrast, Old Earth creationists generally accept the scientific consensus that the universe is approximately 13.8 billion years old. While they believe in the accuracy of modern dating methods, they still reject the notion of biological evolution as described by Darwin. Both Young Earth and Old Earth creationists hold to the inspiration and authority of the Bible as the word of God, though they differ significantly on how to interpret the early chapters of Genesis.

The New Testament presents a different picture of God, emphasizing his loving relationship with humanity through the person of Jesus Christ. Jesus is never portrayed as vengeful but rather as a compassionate savior who desires to reconcile people to God. It is interesting to note that Jesus never used the name Yahweh explicitly in his teachings, instead often referring to God as "Father" (Abba). Jesus' teachings, recorded primarily in the Gospels, emphasize forgiveness, love, and mercy, contrasting with some of the Old Testament's more judicial portrayals.

Ancient Mesopotamian texts contain references to a divine triad consisting of the gods Anu, Enki, and Enlil. According to these stories, Anu and Enlil were responsible for the creation of the heavens and the earth, while Enki created human beings with the help of his consort, Ninhursag or Ninhursaja, who is said to have engaged in a series of experiments to perfect the human form. These mythologies predate the biblical texts and show parallels that have fascinated comparative religion scholars.

The King James Version (KJV) of the Bible, first published in 1611, is the oldest authorized English translation of the Christian scriptures. In the KJV, God instructs Adam and Eve to "replenish" the earth (Genesis 1:28), which some interpret as evidence that there were humans or other beings on earth prior to Adam and Eve. More recent translations of the Bible have replaced the word "replenish" with "fill," significantly altering the meaning of the text.

The Hebrew word male' (מלא) used here means "to fill" or "to be full," and the change in translation reflects evolving understanding of ancient Hebrew. Almost all quotations from the Bible in this book are taken from the KJV and are presented in their original language.

Chapter 30
Anunnaki, God's Creatures and Planets

Given that modern humans were able to interbreed with other hominid species, such as Neanderthals and Denisovans, these archaic human populations had the same number of chromosomes as we do. DNA evidence suggests that the interbreeding between modern humans and Neanderthals occurred approximately 60,000 years ago, while the mixing with Denisovans took place around 50,000 years ago.

The concept of Mitochondrial Eve refers to the most recent common matrilineal ancestor of all living humans. By tracing the DNA passed down from mother to daughter, scientists have determined that this woman likely lived in southern Africa between 150,000 and 200,000 years ago. This finding supports the "Out of Africa" theory, which proposes that modern humans originated in Africa and then migrated to other parts of the world.

The oldest known writings in the world were discovered in Mesopotamia, the birthplace of the biblical figures Abraham and Sarah. Millions of clay tablets called cuneiforms have been found in this region, containing stories of Anunnaki gods, demigods, and everyday life dating back thousands of years before the composition of the Bible. The sheer volume of these written records attests to the level of sophistication and organization achieved by the ancient Mesopotamian civilizations.

According to the Bible, God created Adam and Eve to serve as caretakers of the Garden of Eden. He instructed them to "replenish" the earth, which some interpret as evidence that there were already humans or other beings living on earth before Adam and Eve.

South Africa's Table Mountain, with its flat top that appears to have been artificially leveled, is a striking geological formation. Having personally visited Table Mountain, I can attest to its impressive height and the stunning views it offers of the surrounding landscape.

The exact origin of the creatures that were genetically engineered by the Anunnaki to serve as a labor force remains a mystery. It seems improbable that these beings could have evolved naturally, given the complexity of the genetic code and the intricate processes required for the development of higher life forms. One possibility is that God created these pre-human creatures and that they were later modified by the Anunnaki to suit their purposes. Another theory is that the Anunnaki themselves were created by God, perhaps on another planet, such as Nibiru, that has yet to be found.

According to NASA, "In January 2016, California Institute of Technology (Caltech) astronomers Konstantin Batygin and Mike Brown announced research that provided evidence for a planet about 1.5 times the size of Earth in the outer solar system." The researchers called this unknown body "Planet Nine." Others have called it Planet X for centuries. NASA said, "It could help explain:

1. Why long-period objects in the Kuiper Belt are, on average, tilted by about 20 degrees with respect to the plane within which the planets orbit the Sun

2. Why these long-period orbits cluster in their orientations

3. Why the solar system hosts a distant population of highly inclined trans-Neptunian bodies

4. The existence of objects that reside *between* the giant planets and orbit the Sun in a retrograde direction

5. The persistence of long-period Kuiper Belt objects whose orbits cross the orbit of Neptune

It could also make our solar system seem a little more "normal." Surveys of planets around other stars in our galaxy have found the most common types to be "super-Earths" and their cousins bigger than Earth but smaller than Neptune. Yet none of this kind exists in our solar system. Planet Nine would help fill that gap. In the hunt for Planet 9, scientists will use the Rubin Observatory on top of Cerro Pachón, a mountain in Northern Chile. The observatory is expected to begin operations in 2025. It will conduct a 10-year survey of the Southern Hemisphere sky to help answer some of astronomers' biggest questions about the universe."

According to the Sumerian clay tablets, Enki, the Sumerian god of wisdom, water, and creation, was believed to have played a central role in the development of human civilization. He was often depicted as the son of Anu, the sky god, and was married to the goddess Ninhursag, with whom he had a son named Marduk. Enki was associated with the city of Eridu, one of the oldest known settlements in Mesopotamia.

The Sumerian civilization flourished in the region now known as Iraq, reaching its peak around 4500 BCE. The Sumerians developed a complex system of writing, architecture, art, and scientific knowledge. Their religious beliefs revolved around a pantheon of gods and goddesses who were believed to have direct influence over human

219

affairs. According to Sumerian texts, humans and gods coexisted, with humans serving as laborers and servants to divine beings. The Sumerians were intelligent people who invented many useful things that we take for granted today. Why did they record that they were laborers and servants? Creation myths describing the origins of the world and the formation of human beings were recorded on clay tablets as early as 5000 BCE.

Sumerian texts also describe the creation of a race of hybrid beings formed from the genetic material of the Anunnaki gods and a pre-existing hominid species. These early humans were said to have been created to work as miners, extracting gold and other precious metals from the earth. Interestingly, the tablets suggest that these first humans were sterile and unable to reproduce on their own. This detail is significant because it implies that these beings would not have been able to pass on their genes to future generations, meaning that they could not have been the ancestors of modern humans. The first writings in the world, however, attest to their existence in ancient times.

The idea of a sterile hybrid species is not without precedent in the natural world. The mule, for example, is a cross between a male donkey and a female horse. While mules are known for their strength, endurance, and intelligence, they are unable to produce offspring of their own. In this way, the first human-Anunnaki hybrids may have been similar to mules, possessing desirable traits but lacking the ability to create a self-sustaining population. The mining company I worked for in the past, U.S. Borax, is famous for its twenty-mule team, which was used to transport borax out of Death Valley in the late 19th century. George Washington, who was the first US President, recognized mules as hardy and reliable work animals and also bred them.

In the Sumerian pantheon, Enki was part of a triad of deities who were responsible for the realms of heaven, earth, and the waters. This triad consisted of Anu, the god of the sky, Enlil, the god of the earth, and Enki, the god of wisdom and the subterranean waters. Enki's mother, Nammu, was a primordial goddess who was believed to have given birth to the heavens and the earth. His consort, Ninhursag, was known by many names and titles, including Ninmah and Damgalnuna. Together, Enki and Ninhursag were believed to have created a wide variety of plants, animals, and other living beings.

Observing God's creation, you will notice that every creature has a unique characteristic. Dolphins have sonar; bats, except fruit bats, have radar; an ant can lift 20 times its weight; a flea can jump 200 times its size. This would be equivalent to a 6 feet 6 inches tall person leaping a quarter mile. A mosquito can maneuver better than any sophisticated aircraft or helicopter.

During reproduction, the average octopus lays 100,000 to 500,000 eggs. Several types of trees can live for over five thousand years. Understanding the biology and behavior of other organisms is clearly beneficial to human health. Without bees pollinating plants and making honey, we would have less food to eat. Life fascinates me.

A hummingbird is an animal that glorifies God. It is the only bird able to fly backward and around in virtually any direction due to its wings' ability to rotate in a full circle and can flap 50 to 200 times per second. Its heart can beat up to 1,200 times per minute. The hummingbird is the smallest bird in the world and has the largest brain per body size among birds. They also have the largest heart per body size among all animals. A hummingbird must consume approximately 2 times its weight daily and take an average of 250 breaths per minute at rest.

I don't think it evolved from a dinosaur, even though science says that birds are descendants of dinosaurs. The Hummingbird drawing is the most well-known of all of the geoglyphs among the famous Nazca Lines in Peru!

In addition to their large, multifaceted compound eyes, adult dragonflies also possess two pairs of strong, transparent wings, as well as an elongated body, making them the kings of flight maneuverability. In addition to having nearly 24,000 ommatidia in their eyes, adult dragonflies are able to control their four wings independently, allowing them to fly in any direction, hover, and perform acrobatics. In contrast to most flying insects, dragonflies beat their wings up and down instead of backward and forwards.

It was a pleasure to see rare birds and insects in Taman Negara, the world's oldest tropical rainforest with 130 million years of history in Malaysia. In addition, I enjoyed Singapore's Aviary, which is now Asia's largest bird park and includes more than 3,500 birds.

Also, I enjoyed hiking in Puerto Rico's El Yunque tropical rainforest. The animals were much larger in the past due to warmer temperatures, moist environments, and more oxygen on our planet. Earth's air contained 30-35% oxygen during the Carboniferous and Permian periods, as compared to 21% today.

All other planets in our solar system spin anticlockwise on their axes and orbit the Sun in a counterclockwise direction, except Venus, which spins clockwise on its axis. Uranus is the only planet in the solar system that rotates on its side as it orbits the sun. Venus and Uranus appear to have been perturbed by the gravitational pull of or collision with something powerful. However, we do not know what occurred and when. Did Venus and Uranus have encounters with Planet Nine or Planet X in the past?

Because Neptune is 30 times farther from the sun than Earth, it receives less heat and light, but it radiates a great deal more heat than it absorbs. Winds on Neptune can reach 1,500 miles per hour (2,400 km/h). It is unknown how much Neptune has cooled over its lifetime. Hurricanes on Earth with winds exceeding 157 miles per hour (252 km/h) are considered category 5, the highest level of a hurricane that can cause catastrophic damage.

The gravity of our relatively large moon is responsible for the flow of tides, which carry heat from Earth's equator to the poles and is a crucial part of life on Earth. The climate of Earth would be very different without the moon, and some plants and animals living today could disappear if the moon did not exist. Unless the Moon stabilizes Earth's tilt, the tilt could vary wildly. It would move from having no tilt, which means no seasons, to having a large tilt, which means severe weather.

Chapter 31
Personal Reflections and Beliefs

As I reflect on my own life and the many challenges and opportunities I have encountered along the way, I am struck by the incredible diversity and complexity of the human experience. From my childhood in Burma to my career in the United States, I have had the privilege of meeting and learning from people from all walks of life, each with their own unique perspectives, beliefs, and values.

One of the things that I have come to appreciate most deeply is the power of human connection and the importance of building bridges across divides. Whether it is through shared meals, cultural exchange, or simply taking the time to listen and understand one another's stories, I believe that we have so much to gain by opening ourselves up to the experiences and insights of others.

In fact, research by the Harvard Study of Adult Development one of the longest-running studies on human well-being has shown that strong relationships are among the most consistent predictors of long-term happiness and health.

This is particularly true when it comes to matters of faith and belief. Growing up in a predominantly Buddhist country, I had the opportunity to learn about and appreciate the teachings of the Buddha from a young age. At the same time, I was exposed to a wide range of other religious traditions, from the animist beliefs of my ancestors to the teachings of Confucius and the values of Christianity.

Rather than seeing these different belief systems as contradictory or incompatible, I have come to see them, if possible, as mostly different facets of the same underlying truth the search for meaning, purpose, and connection in a complex and often confusing world.

According to comparative religion scholars such as Huston Smith and Karen Armstrong, many world religions share core ethical teachings such as compassion, humility, and the pursuit of truth, despite theological differences. Whether we call it God, Dharma, or simply the universe, I believe that there is a fundamental unity that underlies all of our diverse experiences and perspectives.

Of course, this does not mean that all beliefs are equally valid or that we should accept everything without question. As someone who values reason, evidence, and critical thinking, I believe that it is important to approach all claims and ideas with a healthy dose of skepticism and to be willing to change our minds in light of new information or arguments.

Jesus said, "I am the way, the truth, and the life." Just as climbers ascend Mount Everest in a single file at the top from the Nepal side, we must be sure to understand that what Jesus said is true. This verse, found in John 14:6, has been a cornerstone of Christian theology, expressing an exclusive truth claim central to Christian belief. However, it has also inspired deep theological dialogue about pluralism and interfaith understanding.

At the same time, I have come to recognize that there are many things in life that cannot be fully understood or explained through reason alone. Love, beauty, and the ineffable sense of wonder and awe that we feel in the face of the universe are all experiences that transcend the boundaries of logic and language.

Psychologists and neuroscientists, including Dr. Andrew Newberg, have studied these "mystical experiences" and found that spiritual or transcendent moments can activate parts of the brain associated with emotional insight, empathy, and creativity.

In this sense, I believe that faith and reason are not necessarily opposed but rather complementary aspects of the human experience. While science and rationality can help us understand the workings of the natural world and solve practical problems, it is through faith and intuition that we are able to connect with something greater than ourselves and to find meaning and purpose in our lives. Albert Einstein himself once noted that "science without religion is lame, religion without science is blind," reflecting the possibility of integration rather than opposition.

This is not to say that faith is always easy or straightforward. As someone who has grappled with questions of belief and meaning throughout my life, I know firsthand how challenging it can be to find a true sense of purpose and direction in a world that often seems chaotic and unpredictable.

But I have also come to believe that it is through this struggle through the willingness to ask difficult questions, to confront our own doubts and fears, and to remain open to new possibilities that we are able to grow and evolve as individuals and as a society.

In my own life, I have found that one of the most powerful ways to cultivate this sense of openness and curiosity is through travel and exposure to different cultures and ways of life. From the bustling streets of Tokyo to the ancient ruins of Rome, I have had the opportunity to see the world through the eyes of others and to gain a deeper appreciation for the richness and diversity of the human experience.

According to a 2010 study published in the Journal of Personality and Social Psychology, individuals who travel and live abroad develop greater cognitive flexibility, enhanced creativity, and a more nuanced understanding of cultural norms.

One of the most memorable experiences of my life was attending the World Expo in Vancouver with my family. Seeing the incredible diversity of cultures, innovations, and ideas on display was a powerful reminder of the incredible potential of human creativity and collaboration.

Expo 86, held in Vancouver, British Columbia, was attended by over 22 million people and featured 54 nations. Its theme, "World in Motion – World in Touch," emphasized technological innovation and global connectivity. It showcased developments in communication, transportation, and global culture leaving a lasting impression on a generation of attendees.

At the same time, I have also been deeply moved by the simple acts of kindness and generosity that I have witnessed in my own community and around the world. From the volunteers who give their time and energy to help those in need to the everyday heroes who stand up for what is right and just, I am constantly inspired by the power of ordinary people to make a difference in the world.

Studies in positive psychology, particularly the work of Dr. Martin Seligman and Dr. Sonja Lyubomirsky, show that acts of kindness not only benefit recipients but also significantly boost the mental and physical well-being of those who give.

As I look to the future, I am filled with both hope and concern. On one hand, I see incredible opportunities for progress and positive change driven by advances in science, technology, and human understanding. From the development of clean energy and sustainable agriculture to the eradication of poverty and disease, I believe that we have the knowledge and the tools to create a better world for all.

Initiatives such as the United Nations' 17 Sustainable Development Goals (SDGs) provide a global framework for tackling poverty, inequality, and climate change by 2030, highlighting how international collaboration and innovation can lead to tangible progress.

At the same time, I am deeply troubled by the many challenges and threats that we face as a global community. From the existential threat of climate change to the rise of authoritarianism and the erosion of democratic norms, I worry that we are losing sight of the values and principles that have guided human progress for generations.

According to the 2024 Intergovernmental Panel on Climate Change (IPCC) report, without urgent global action, average temperatures could surpass the 1.5°C threshold within the next two decades triggering severe environmental and humanitarian consequences. Simultaneously, organizations like Freedom House have documented a decline in global democratic freedoms for nearly two decades.

But I also know that we have faced similar challenges before and have found ways to overcome them through the power of human ingenuity, compassion, and resilience. Whether it is through the development of new technologies, the forging of new alliances and partnerships, or simply the willingness to stand up for what is right and just, I believe that we have the capacity to create a better future for ourselves and for generations to come. The successful global effort to eradicate smallpox by 1980 achieved through international cooperation and innovation is a powerful historical example of what humanity can accomplish when united by purpose.

Ultimately, I believe that the key to unlocking this potential lies in our ability to cultivate a sense of empathy, understanding, and connection with one another. By listening to each other's stories, learning from each other's experiences, and working together towards common goals, we can build a world that is more just, compassionate, and resilient.

Empathy, now increasingly recognized as a vital social skill by institutions like the Harvard Graduate School of Education, is foundational to resolving conflict and fostering cooperative societies, especially in diverse, interconnected communities.

This is not a task that any one person or group can accomplish alone. It will require the efforts and contributions of people from all walks of life, from all corners of the globe. But I am convinced that, by working together and remaining true to our highest values and aspirations, we can create a future that is worthy of the incredible potential of the human spirit.

Chapter 32
Signs of the End Times

As I reflect on my own journey and the many twists and turns that have brought me to where I am today, I am filled with a deep sense of gratitude for the people and experiences that have shaped my life. From my parents and family to my teachers, mentors, and friends, I have been blessed with an incredible network of support and guidance along the way.

I am also grateful for the many challenges and obstacles that I have faced, for it is through these struggles that I have learned and grown the most. Whether it was navigating the complexities of cultural differences, overcoming personal and professional setbacks, or simply learning to trust in my own strengths and abilities, each challenge has taught me valuable lessons and helped me to become a stronger, more resilient person.

Resilience, as defined in psychological studies led by Dr. Ann Masten, is not a rare trait but a common capacity that can be nurtured through meaningful relationships, positive outlooks, and adaptive coping strategies.

Looking ahead, I know that there will be many more challenges and opportunities to come. But I am excited to embrace them with an open mind and a willing heart, knowing that each new experience will bring with it the chance to learn, grow, and make a positive impact in the world.

And so, to all those who are reading these words, I offer

this simple message: never stop learning, never stop growing, and never stop believing in the power of the human spirit to overcome even the greatest of obstacles. For it is through our shared humanity, our common hopes and dreams, and our unwavering commitment to building a better world that we will find the strength and the courage to create a future that is truly worthy of our highest aspirations.

Author and Holocaust survivor Viktor Frankl once wrote, "When we are no longer able to change a situation, we are challenged to change ourselves." This spirit of adaptation, purpose, and endurance is what I hope we can all carry forward into the future.

We are living in the most exciting and challenging time in human history. Focus your attention on the things that matter most to you. Nobody knows the equation for love, but everyone knows love when they have experienced it for themselves.

While love continues to defy formal definition, modern neuroscience through the work of Dr. Helen Fisher and others has mapped some of its biological underpinnings, linking it to neural pathways related to attachment, empathy, and reward. Yet its true meaning remains deeply personal and universally profound.

Thank you for taking the time to read and reflect on these words. I hope that they have offered some insight, inspiration, or simply a moment of connection in a world that can often feel overwhelming and uncertain. May we all find the wisdom, compassion, and resilience to navigate the challenges ahead and create a future that is filled with hope, joy, and endless possibilities.

Section 5: End Times Prophecy

Chapter 32
Signs of the End Times

The Bible provides a prophetic roadmap of events that will unfold before the return of Jesus Christ. These signs of the end times include global conflict, disease outbreaks, and a decline in genuine faith. Perhaps most troubling is the rise of an apostate church—one that strays from divine truth and embraces false doctrines.

While these prophecies may seem alarming, Scripture reminds us that God remains sovereign over all history. In light of this, our response should be to deepen our trust in Him and to continue demonstrating love—for God and for one another.

Perilous Times Foretold

Paul's second letter to Timothy offers a sobering description of the last days:

"This know also, that in the last days perilous times shall come. For men shall be lovers of their own selves, covetous, boasters, proud, blasphemers, disobedient to parents, unthankful, unholy, Without natural affection, trucebreakers, false accusers, incontinent, fierce, despisers of those that are good, Traitors, heady, high-minded, lovers of pleasures more than lovers of God; Ever learning, and never able to come to the knowledge of the truth."

(2 Timothy 3:1–7)

This portrait mirrors much of what we see today—moral confusion, pride, and rebellion against truth.

233

Christianity and Islam: Irreconcilable Doctrines

The Qur'an denies both the crucifixion and resurrection of Jesus—central tenets of the Christian faith. Islam teaches that Jesus was not crucified but that someone else was made to look like him. In contrast, Christianity affirms that Jesus died on the cross as the ultimate sacrifice for humanity's sins and rose again to conquer death, just as He foretold.

This fundamental contradiction raises serious questions: How can Pope Francis claim these faiths are equal or compatible? How many Catholic churches have held Ramadan celebrations, and why? Some believe Pope Francis played a role in the resignation of Pope Benedict XVI—an event marked, symbolically, by two lightning strikes hitting the Vatican on that very day.

The Irreplaceable Truth of Jesus Christ

As of 2023, there were 2.3 billion Christians, comprising 31.5% of the global population. Jesus Christ's ministry spanned just three years, yet He remains the most impactful figure in human history. Despite scholarly debate, historical evidence overwhelmingly supports the reality of His life, crucifixion, and the explosion of the early church based on His resurrection.

While Christians and Muslims can and do live in peace, their religious foundations are irreconcilable. There is only one truth. I have had meaningful and respectful conversations with Muslim neighbors in Saratoga and even shared a Ramadan meal with Muslims in Egypt. But theological honesty requires acknowledging that peaceful coexistence does not equal doctrinal agreement.

The Jewish Temple and the Sacred Rock

The Jewish people built their first temple during the reign of King Solomon, son of King David, completing it in 957 BCE. It was constructed around the rock where Abraham prepared to sacrifice his son Isaac, according to the Torah.

The Qur'an, written between 610–632 CE, claims it was Ishmael—not Isaac—who was to be sacrificed. But if that were true, it seems unlikely that the Jewish temple would have been built on that specific site. This discrepancy underscores the theological and historical tension over this sacred ground.

The Rock of Contention

The Prophet Muhammad is traditionally believed to have ascended into heaven from the same rock. Today, this relatively small outcrop in Jerusalem is the most contested piece of land on Earth, holding the potential to ignite a global conflict—possibly even World War III.

Buddhism and the Pursuit of Nirvana

Buddhism teaches that human effort can lead to Nirvana, an enlightened state free from suffering and rebirth. Unlike Christianity, Buddhism does not require belief in a personal God. Instead, followers are encouraged to rely on themselves—a stark contrast to the Christian call to rely wholly on God's grace through Jesus Christ.

Resurrection and the Call to Love

Jesus said, "Love one another as I have loved you." That command, given two thousand years ago, remains radically relevant today. He also urged His followers to pray for those who persecute them and to forgive freely.

As Revelation 22:10–11 proclaims:

"And he saith unto me, Seal not the sayings of the prophecy of this book: for the time is at hand. He that is unjust, let him be unjust still: and he which is filthy, let him be filthy still: and he that is righteous, let him be righteous still: and he that is holy, let him be holy still."

These words remind us that each person must choose whom they will serve, especially as the final days draw near.

End-Time Deception and Emerging Technologies

Scripture warns repeatedly of increasing deception in the last days. Today, that deception is amplified by biased media, manipulated narratives, and the rise of artificial intelligence. With AI and artificial general intelligence (AGI) advancing rapidly, it is becoming more difficult to distinguish between truth and fabrication.

This mirrors biblical warnings: "Even the elect could be deceived, if that were possible" (Matthew 24:24). The very tools meant to inform may, ironically, become instruments of mass deception, reinforcing the urgency of staying rooted in God's Word.

Natural Disasters and the Fragility of Creation

Natural disasters are also highlighted in biblical end-times prophecies. We can expect to see an uptick in devastating events such as wildfires, prolonged droughts, catastrophic floods, global pandemics, powerful earthquakes, and violent volcanic eruptions. These events serve as stark reminders of the earth's fragility and humanity's profound need for divine protection and intervention.

The Rise of the Beasts and Spiritual Deception

The Bible speaks of two metaphorical beasts that will arise in the last days, aiming to mislead humanity and draw people away from faith in God. These prophetic passages underscore the urgent need for spiritual discernment in these perilous times. Believers must exercise caution regarding whom they trust, testing every teaching against the authority of Scripture.

Even in the face of deception, we take comfort in knowing that Jesus Christ's power far surpasses that of any earthly or spiritual adversary.

The Mark of the Beast and the Coming Economic System

A pivotal end-time prophecy involves the implementation of a "mark of the beast", a sign required to participate in commercial transactions. This mark is described in Revelation 13, and while its form may become ubiquitous, the greater concern lies in its use as a tool for controlling personal freedom and economic access.

"And he causeth all, both small and great, rich and poor, free and bond, to receive a mark in their right hand, or in their foreheads: And that no man might buy or sell, save he that had the mark, or the name of the beast, or the number of his name. Here is wisdom. Let him that hath understanding count the number of the beast: for it is the number of a man; and his number is Six hundred threescore and six." (Revelation 13:16–18)

Revelation 13:11–12 further describes the second beast:

"And I beheld another beast coming up out of the earth; and he had two horns like a lamb, and he spake as a dragon. And he exerciseth all the power of the first beast before him, and causeth the earth and them which dwell therein to worship the first beast, whose deadly wound was healed."

These cryptic verses raise important questions about the identities of these entities and the nature of the "healed wound", suggesting a political or spiritual figure whose return to power will be globally recognized and worshipped.

As we witness the rise of digital currencies, real-time facial recognition, biometric scanning, location tracking, and centralized QR-based payment systems, the infrastructure for a system of global surveillance and control already exists. Authorities could potentially block individuals from buying or selling, track associations, and monitor dissent—all in real time.

Sober Watchfulness Anchored in Hope

While these developments are sobering, the Bible calls us to approach prophecy with both alertness and hope. These signs paint a picture of mounting global tension, but they also affirm God's ultimate control over history and His promise to redeem those who remain faithful.

Fulfillment of prophecy affirms the divine inspiration of Scripture and strengthens our faith. As global events align with ancient predictions, we are called to respond not with fear, but with spiritual preparation.

Spiritual Readiness in a Time of Testing

Believers are urged to maintain a constant state of **spiritual readiness**. This includes:

- **Cultivating a strong prayer life**
- **Diligent study of Scripture**
- **Active participation in a community of faith**
- **Standing firm in biblical convictions**, even under **societal pressure or persecution**

We are not called merely to survive, but to live with courage, compassion, and conviction—bearing witness to God's truth in a darkening world.

The Apostate Church and Compromised Faith

One of the more concerning signs of the end times is the rise of the Apostate Church—a religious institution that outwardly resembles Christianity but has inwardly departed from the truth. Such institutions may prioritize worldly acceptance over biblical fidelity, compromising core doctrines to align with cultural trends. Believers must remain discerning, evaluating all teachings and practices against the unchanging standard of God's Word.

False Prophets, Signs, and Deception

The proliferation of false prophets and teachers is another unmistakable sign. These individuals may perform signs and wonders, appearing convincing and trustworthy. Jesus warned that their deception would be so potent that, if it were possible, even the elect could be deceived (Matthew 24:24). This underscores the need for deep biblical knowledge and continual discernment through the Holy Spirit.

Globalism and the Coming World Order

The Bible also points to a future global government and economic system that will exert control over all buying and selling. Current movements toward globalization, centralized authority, and international economic regulation provide a possible pathway for this system to emerge.

Environmental Shifts and the Birth Pains

Scripture refers to "birth pains"—environmental and social upheavals that precede the end. These include:

- **Increased intensity and frequency of natural disasters**
- **Extreme weather events and climate shifts**
- **Widespread ecological degradation**

These changes reflect the groaning of creation and point toward the coming redemption described in Romans 8:22–23, "For we know that the whole creation groaneth and travaileth in pain together until now. And not only they, but ourselves also, which have the firstfruits of the Spirit, even we ourselves groan within ourselves, waiting for the adoption, to wit, the redemption of our body."

Technology, Surveillance, and Control

While technology offers many benefits, it also presents unique challenges in light of prophecy. The development of:

- **Advanced surveillance systems**
- **Implantable identification technologies**
- **Complete digitization of personal and financial data**

This raises serious concerns about how authoritarian regimes could wield these tools for global control, as predicted in Revelation.

Hope Anchored in Christ's Return

Despite all these signs and warnings, the Bible consistently encourages us to remain hopeful. The same prophecies that foretell tribulation also promise victory—the return of Christ, the defeat of evil, and the establishment of God's eternal kingdom.

We are not merely passive observers—we are called to be Christ's ambassadors. As we await His return, we must share His truth, embody His love, and live with steadfast hope, no matter how chaotic the world becomes.

Are we witnessing the beginning of the 3-1/2-year tribulation revealed in the Book of Revelation? The US is the Second Beast in Revelation. The Trump administration, whose term will end in about 3-1/2-years, has

1. Caused or is causing 300,000 deaths by reducing or eliminating USAIDS. According to reports from NPR, Democracy Now!, and The Times, modeling from Boston University estimates that reductions in funding for the U.S. Agency for International Development (USAID) may have led to nearly 300,000 deaths.
2. Attacked Iran, focusing on nuclear facilities, using some 125 military aircrafts, including 7 B-2 stealth bombers and fourth- and fifth-generation fighters, 14 30,000 pounds deep penetrating bombs, 75 precision guided weapons and more than two dozen Tomahawk missiles fired from a nuclear submarine primarily focused on surface infrastructures. In

joining the Israel-Iran war, US made deliberate choices to violate the US Constitution and UN Charter, and acted lawlessly to risk overt and covert retaliatory attacks that could lead to WWIII for generations to come.

3. Accelerated the impending global climate change disaster by withdrawing from the 2015 Paris Accords adopted by 195 Parties at the UN Climate Change Conference (COP21) in Paris, France, on December 12, 2015. It entered into force on November 4, 2016.

4. Withdrawn from World Health Organization (WHO) and jeopardized critical health initiatives, weakened international cooperation and undermined efforts to address pressing global challenges.

5. Increased chaos and anxiety globally by imposing unfair tariffs that would have to be paid mostly by Americans and billions of people globally could suffer due to disruption in trade and other non-economic activities.

6. Has taken authoritarian actions resulting in the "No Kings" movement that has produced thousands of protests nationwide and globally involving millions of people largely against the policies and actions of Donald Trump's second presidency.

7. Deployed the National Guard troops and Marines to confront protesters in Los Angeles, where protests against immigration raids by the Trump administration led to clashes in the streets.

8. Revoked $11 billion in funding for addiction and mental health care. More than 209,000 Americans die each year from alcohol, suicide and drug overdoses. These conditions account for an estimated $700 billion annually.

9. Conspired with the Netanyahu administration to enable the Israel IDF to bomb Iran and assassinate

their leaders and scientists, resulting in severe retaliation by Iran. This could easily lead to WWIII.

10. Allowed the genocide and ethnic cleansing in Gaza to continue and more than 55,000 Palestinians, mostly women and children, have died and 1.8 million displaced people don't have enough food, water, and medicine due to Israel blockades.

11. Tried to end the Ukraine-Russia war, but failed so far, and the intensity of the damages done by both sides is increasing. The EU, NATO, and G-7 don't want to end this war, which has caused more than a million deaths or injured and turned more than 6.9 million Ukrainians into refugees in other countries, primarily in Europe. As of mid-2025, approximately 3.7 million people are internally displaced within Ukraine.

12. Demanded NATO partners to increase defense spending to 5% on the backs of the European citizens, who will see continued degradation in their quality of life.

13. Taken actions to legitimize cryptocurrency, which is backed by nothing. In time, this Ponzi scheme will become self-evident. Like the creation of the Federal Reserve that facilitated perpetual wars resulting in over 100 million deaths, cryptocurrency can do a lot of damage to billions of people around the world.

14. Fired many scientists who have contributed to improving health for millions of people and pushing to cut assistance to those who can least afford healthcare and food.

15. Ignored homelessness, which is fast becoming a crisis nationally. In 2024, there were approximately 771,400 homeless individuals in the country, an increase of 118,300. Many famous tourist destinations such as the Fishermen's Wharf in San

Francisco have turned into abandoned neighborhoods.

16. Made the US economy worse, people are suffering and some countries refuse to accept "US Dollar" as payment for their goods. Annualized GDP growth for the first quarter of 2025 was -0.3%, a significant drop from the previous quarter's 2.4%.

17. In 2025, the U.S. dollar is down almost 10% year-to-date against the euro, down more than 8% against the Mexican peso, and down 8% against the Japanese yen. The US government is projected to spend over $1 trillion annually on interest payments for its national debt of more than $37 trillion.

18. Taken actions to limit "Free Speech," especially on university campuses and cut billions in funding to universities such as Harvard, which is older than the US.

19. Governed through executive orders, issuing 142 orders in the first 100 days. These orders are about 5 times more than most presidents since Franklin D. Roosevelt issued 99 to rescue the nation from the Great Depression.

20. Caused significant decline in truth. According to The Washington Post, Trump made more than 30,000 false and misleading claims in his first term and it hasn't stopped.

21. Selected his cabinet based upon loyalty to Trump and not merit. What would be the consequence of this approach?

22. Devalued and politicized the justice system significantly. President Trump is the first convicted felon ever to become president and he has regularly attacked the courts, and wielded investigative powers to target his opponents.

23. Used the powers of government against major institutions Trump dislikes, like law firms,

universities and the media, in ways that are often transparently political.

24. Instilled a culture of fear that leads to violence.
25. Caused a decline in trust by allies and not only adversaries. This could accelerate the decline of the US dollar as the global currency.

Politics, Prophecy, and the Call for Spiritual Vigilance

I did not vote in the 2024 presidential election because I believed I would regret casting a ballot for either Donald Trump or Kamala Harris. Instead, I pray that God would protect President Trump and guide him to make decisions that promote justice, humility, and global peace.

Whether one believes in the divine inspiration of the Bible or not, there is timeless practical wisdom in its pages. America's political obsession with containing the rise of China, for example, is not just geopolitical—it reflects the age-old sin of envy, one of the seven deadly sins.

The Bible tells us that envy triggered the first recorded murder, when Cain killed his brother Abel. It was also envy that drove Satan to rebel against God, ultimately leading to his fall from heaven to Earth. This spiritual condition continues to shape global conflict and policy today.

Historical Interventions and the Cost of Global Leadership

A brief survey of history illustrates how power and envy have long driven international conflicts, often under the banner of humanitarianism. NATO, formed in 1949, did not conduct its first military intervention until 1995—46 years after its inception. That action, heavily influenced by President Bill Clinton, marked the beginning of a series of

U.S.-led interventions without UN authorization, raising serious questions of international legality and moral consistency.

In March 1999, the U.S.-led NATO bombing of Yugoslavia was justified by the Kosovo crisis. The 78-day campaign targeted military and civilian infrastructure alike—oil refineries, factories, and even television stations—without the approval of the UN Security Council or General Assembly. It resulted in the deaths of 2,500 people and displaced over a million civilians.

Meanwhile, during this period, President Clinton admitted to an affair with Monica Lewinsky—an act that violated the sanctity of the Oval Office. Just three days after this public admission, on August 20, 1998, Clinton authorized Tomahawk missile strikes in Afghanistan and Sudan, one of which reportedly killed Osama bin Laden's daughter. The political timing of that strike raised questions that history has not forgotten.

The War on Terror and Its Fallout

In 2001, Osama bin Laden orchestrated the 9/11 attacks, which brought down the World Trade Center towers, damaged the Pentagon, and resulted in 2,977 deaths—the deadliest terrorist attack in history. The U.S. response—the global War on Terror—spanned decades, cost millions of lives, and fundamentally reshaped international law, surveillance, and military engagement.

The Proxy War in Ukraine

Today's Ukraine-Russia conflict is widely seen as a proxy war between NATO and Russia. Its origins trace back to President Clinton's decision to expand NATO eastward in

1999, a move that broke earlier promises made to Russian leaders in exchange for agreeing to the reunification of Germany. This act, though viewed as strategic by some, is seen by others as the seed of provocation that led to the unfolding catastrophe—a war that has killed or injured over a million people.

Gaza, Genocide, and Global Hypocrisy

The current war in Gaza has been labeled by numerous international watchdogs—including Amnesty International and UN Special Rapporteurs—as a genocidal campaign. These organizations cite large-scale civilian deaths, the destruction of critical infrastructure, and the targeting of healthcare facilities. In many documented cases, there is no evidence that these strikes had legitimate military objectives.

In contrast to its intervention in Yugoslavia, NATO has taken no action in Gaza, despite the scale and intensity of the violence far surpassing that in Kosovo. This disparity in response raises difficult moral questions about selective justice, geopolitical bias, and who gets to define humanitarian crisis.

June 2025: The Israel-Iran War

In June 2025, Israel bombed Iran, killing at least 865 people and wounding more than 3,300. These attacks were carried out without UN authorization, violating international law. The United States subsequently joined the conflict, launching its own strikes against Iran—again bypassing both the UN Security Council and the U.S. Constitution, which requires Congressional approval for acts of war.

This act mirrors earlier unilateral actions and reflects a growing pattern of executive overreach, contributing to the destabilization of already fragile global relationships.

A Spiritual Lens on Global Events

Taken together, these modern events align eerily with biblical prophecy. The escalation of global warfare, selective justice, technological control, and institutional deception are all signs Scripture warns will precede the return of Christ.

While these signs may seem daunting, they serve as a call to spiritual vigilance, deeper faith, and active moral engagement. By understanding these prophecies and their real-world implications, believers can face the days ahead with wisdom, courage, and unshakable trust in God's sovereign plan.

Final Reflection: Living with Hope Amid Turmoil

As we confront political instability, war, and the increasing collapse of moral clarity, we must not lose heart. The believer's hope is not in worldly leaders or political alliances, but in the eternal kingship of Christ. These signs of the end times are not just warnings—they are also reminders of God's control, His justice, and His coming restoration.

"Therefore keep watch, because you do not know on what day your Lord will come." (Matthew 24:42)

Let this moment in history awaken our souls—not just to the fragility of human empires—but also to the enduring strength of divine truth.

Chapter 33
The Rise of Digital Currency

The world is rapidly transitioning away from physical money toward digital forms of currency. This shift marks a paradigm change in global finance, revolutionizing how individuals conduct transactions, store value, and engage with banking systems. While digital currency may seem novel to some, many Americans have already embraced digital banking services, with over 65% of U.S. adults using mobile banking apps as of 2024.

Financial institutions continue to introduce advanced digital tools, including AI-powered budgeting assistants, biometric authentication, and contactless payment systems. These innovations reflect a broader move toward a cashless society, driven by convenience, security, and efficiency.

Digital currencies exist primarily in electronic form and are exchanged through secure computer networks, especially those connected to the Internet. Unlike credit card payments or online transfers tied to fiat currencies, true digital currencies—such as cryptocurrencies and Central Bank Digital Currencies (CBDCs)—never exist in physical form. This distinction makes them uniquely compatible with our increasingly interconnected, cash-optional world.

Central banks in major economies—including Brazil, China, the Eurozone, India, Nigeria, and the United Kingdom—are leading the way in exploring or implementing CBDCs. As of 2024, over 130 countries, representing more than 98% of global GDP, are researching or developing CBDCs, according to the Atlantic Council's GeoEconomics Center.

Chapter 33
The Rise of Digital Currencies

China's digital yuan (e-CNY) has already been launched in pilot programs across dozens of cities, with over 260 million wallets opened and billions in transaction volume. The Bahamas became the first nation to roll out a fully operational CBDC, the Sand Dollar, in 2020, setting a precedent for small economies leveraging digital finance to enhance accessibility.

In the U.S., a pilot initiative led by the Federal Reserve Bank of New York is testing digital dollar tokens for interbank settlements. Participants include major institutions such as BNY Mellon, Citi, HSBC, Mastercard, PNC, TD Bank, Truist, U.S. Bank, and Wells Fargo. This signals strong institutional momentum toward mainstream adoption of digital currency infrastructure.

JPMorgan Chase, meanwhile, has already implemented JPM Coin—its own internal digital token used for wholesale transactions. The system enables real-time, blockchain-based transfers of dollar and euro balances, processing over $1 billion per day as of 2023. While this represents just a small share of JPMorgan's daily $10 trillion flow, it highlights the efficiency gains digital tokens offer in high-volume environments.

However, CBDCs raise major privacy concerns. Unlike cash, which offers anonymity, digital currencies can allow governments to track every transaction. Critics argue this could pave the way for surveillance and limit personal freedom. In authoritarian regimes, programmable money could even be used to restrict purchases or freeze dissenters' accounts. These risks warrant close attention as the U.S. considers its digital dollar roadmap.

The QR code (Quick Response code)—invented by Japan's Denso Wave in 1994 to track auto parts—has now become a staple in digital commerce. It facilitates contactless payments and can store up to 4,000 alphanumeric characters. In countries like China, QR-based payment apps like WeChat Pay and Alipay dominate daily life, processing trillions of dollars in transactions annually.

In a fully digitized economy, it's conceivable that QR codes or similar technologies—possibly embedded in digital IDs or mobile wallets—will become required for purchases. This has sparked debate among civil liberties advocates, who warn that tying personal identification to every transaction could threaten financial autonomy and increase vulnerability to cybercrime.

The global digital currency shift also carries macroeconomic and geopolitical consequences. By digitizing national currencies, central banks may gain unprecedented control over monetary policy, enabling real-time interventions such as direct stimulus payments or programmable interest rates. However, this centralization could reduce the role of commercial banks, forcing a structural evolution in traditional banking.

Moreover, CBDCs have the potential to challenge the U.S. dollar's role as the world's reserve currency. For example, China's efforts to internationalize the digital yuan in cross-border trade—particularly through its Belt and Road Initiative—signal a strategic bid to reduce dependence on dollar-based systems.

At the same time, digital finance could deepen global inequalities. Roughly 2.6 billion people remain offline, and many lack access to smartphones, secure IDs, or digital literacy. If digital currency becomes the primary medium of exchange, those without access risk being excluded from essential services. This raises urgent questions about digital inclusion and infrastructure development.

The environmental impact is another critical issue. While CBDCs are typically energy-efficient, many cryptocurrencies like Bitcoin rely on proof-of-work mining, which consumes immense amounts of electricity. Bitcoin alone is estimated to use more energy annually than the entire country of Argentina. Transitioning to greener blockchain protocols (e.g., proof-of-stake) and regulating crypto mining could help reduce digital finance's carbon footprint.

Cybersecurity is also paramount. As digital assets become more valuable and widely used, they become bigger targets for hackers. High-profile breaches—such as the $600 million Poly Network hack—underscore the risks. Governments and private firms must invest in secure infrastructure and establish clear consumer protections.

Emerging technologies like AI and the Internet of Things (IoT) are accelerating the shift. For instance, IoT-enabled smart devices could trigger automated payments for services like energy or tolls, while AI could manage investment portfolios or detect fraud in real time. These developments offer convenience—but also increase system complexity and the stakes of failure.

On an individual level, navigating the transition to digital finance will require new skills and habits. Financial literacy education—especially around cybersecurity, personal data protection, and digital wallets—will become essential. Tools like biometric authentication and multi-factor verification can enhance security, but they also raise concerns about data sovereignty and surveillance.

In conclusion, the rise of digital currencies represents a transformative moment in human economic history. If implemented thoughtfully, they can improve efficiency, expand financial access, and enable smarter policy responses. But if rushed or poorly regulated, they could increase inequality, infringe on freedoms, and destabilize traditional institutions. Balancing innovation with oversight will be key.

As we enter this new financial era, individuals, businesses, and policymakers must approach digital currencies with open eyes, informed by both the promise and the perils of this powerful technology.

Chapter 34
Global Adoption of QR Codes and Digital Payments

China and India are leading the world in the adoption of QR code payments, with China having the largest number of mobile payment users globally. As of 2024, over 90% of urban Chinese consumers regularly use QR codes for everyday transactions via platforms like WeChat Pay and Alipay, which together process trillions of dollars annually. India's Unified Payments Interface (UPI) has also facilitated a massive surge in QR code payments, particularly through apps like PhonePe, Paytm, and Google Pay India, which are now widely accepted even by street vendors and small businesses.

Several Asia-Pacific nations, including Indonesia, Singapore, Malaysia, Thailand, and the Philippines, are also embracing QR code technology. For example, Singapore's PayNow and Thailand's PromptPay have national QR code standards to ensure interoperability between banks and merchants. It's worth noting that these nations are not predominantly Christian, although Christian communities do exist within them. Many people in these regions may be unfamiliar with or skeptical of Biblical prophecies—particularly those concerning the end times—due to differing cultural and religious perspectives.

The slower adoption of QR codes in the United States was primarily due to early smartphones lacking native QR scanning capabilities, which required users to download third-party applications. However, this barrier began to fade after 2017, when Apple integrated QR code scanning directly into its iPhone camera app (iOS 11), followed by similar changes in Android devices. Since then, QR code usage in the U.S. has increased significantly, particularly in response to COVID-19, which normalized touch-free payments in restaurants, retail, and healthcare.

QR codes have the ability to store more information than traditional barcodes—up to 4,296 alphanumeric characters or 7,089 numeric characters—and their ease of use has driven adoption across various industries, especially in digital payments. The convenience is undeniable: customers scan a code, are directed to a secure payment page, and complete the transaction using options like Apple Pay, Google Pay, Samsung Pay, or linked bank accounts. This process is fast, secure, and eliminates the need for cash or physical credit cards.

However, the rise of QR code payments is not without risks. Phishing attacks and malicious QR codes—referred to as "quishing"—are on the rise. Hackers can embed dangerous URLs in QR codes that, once scanned, can prompt the download of malware or redirect users to counterfeit websites to steal personal and banking information. In 2022, the FBI issued an alert about this growing threat, emphasizing the need for vigilance and encryption in QR-based systems.

QR code payments also depend on a smartphone with a camera and reliable Internet access, which may not be consistently available in all locations. This dependence can present challenges in rural areas or for populations that are technologically underserved. Additionally, older generations or individuals uncomfortable with smartphones may find the shift to QR payments confusing or alienating.

To address such challenges, the U.S. government offers subsidized mobile devices and Internet access to low-income individuals through programs like Lifeline and the

Affordable Connectivity Program. While these initiatives are intended to bridge the digital divide, they also unintentionally accelerate the mass adoption of digital technologies, including QR-based payment platforms, and may hasten the transition away from traditional cash-based transactions.

This global shift toward QR code payments raises profound questions about financial inclusion, accessibility, and ethical oversight. While QR codes simplify and speed up transactions, they may also exclude or marginalize those without smartphones, banking access, or digital literacy. Ensuring that vulnerable populations aren't left behind will be crucial as societies increasingly depend on these systems.

Given the ease and ubiquity of QR code scanning, it is not inconceivable that one day individuals might be asked—or required—to carry a personalized, scannable identifier. This could take the form of a wearable chip, digital ID, or even an invisible ink QR code on the hand or forehead— technologies that are already being explored for secure identification and health tracking purposes. Though speculative, such developments echo concerns raised by theologians and privacy advocates alike.

From a Biblical perspective, these technological advancements bring renewed attention to prophecies in the Book of Revelation concerning the "mark of the beast." The Bible warns that receiving this mark—whether in the hand or forehead—will have eternal spiritual consequences.

Revelation 14:9–11 (KJV) warns:

"If any man worship the beast and his image, and receive his mark in his forehead, or in his hand, The same shall drink of the wine of the wrath of God... and he shall be tormented with fire and brimstone... And the smoke of their torment ascendeth up for ever and ever..."

Revelation 16:1–2 describes divine judgment upon those who accept this mark:

"...there fell a noisome and grievous sore upon the men which had the mark of the beast, and upon them which worshipped his image."

These verses highlight that the decision to accept such a mark—no matter how practical or socially necessary it may seem—has deeper spiritual implications. What may appear as a harmless digital ID or payment method could, in the eyes of believers, represent a submission to systems that oppose God's authority.

As digital payment technologies become more ubiquitous and potentially mandatory for participation in economic life, believers may soon be forced to discern carefully how to engage with these systems. This will involve not just technical or financial decisions, but spiritual ones, grounded in scripture and guided by prayer, community, and conviction.

It's important to note that not all digital payment systems or technological advancements inherently represent the mark of the beast as described in the Book of Revelation. Many of these innovations—such as contactless payments, mobile wallets, and blockchain-based transfers—serve practical purposes and improve convenience and efficiency. However, the accelerating trend toward cashless societies and the integration of biometric and digital identity systems into everyday commerce certainly creates an environment in which such a system could, one day, be implemented.

Chapter 34
Global Adoption of QR Codes and Digital Payments

As we observe these technological and economic transformations, it is crucial for believers to remain informed, spiritually grounded, and discerning. While we cannot know precisely how end-time prophecies will unfold, Scripture urges vigilance and preparedness. Matthew 24:42 reminds us, "Therefore keep watch, because you do not know on what day your Lord will come." This means staying alert, not fearful—prioritizing our allegiance to God over convenience, comfort, or conformity to emerging economic systems.

The adoption of QR codes and digital payments also raises urgent concerns about data privacy, surveillance, and control. Every transaction completed through these systems can be digitally logged, timestamped, and associated with personal identifiers. Companies and governments are increasingly capable of analyzing this data to derive behavioral insights and implement predictive analytics. While this can help prevent fraud, optimize services, and combat illegal activity, it also opens the door to highly invasive forms of monitoring, censorship, and social control—especially in authoritarian regimes or under unchecked corporate influence.

Moreover, the shift to digital payment infrastructures has major implications for monetary policy and global financial stability. Central banks may acquire more precise tools to monitor consumer activity or implement targeted stimulus, such as programmable CBDCs (Central Bank Digital Currencies). However, managing a hybrid system of physical and digital currencies poses technical and regulatory challenges, including risks of cyberattacks, liquidity imbalances, and increased dependency on centralized databases.

The cross-border nature of digital currencies and mobile payment platforms also introduces new complexities for international finance. Interoperability between national digital currencies (such as China's digital yuan, India's e-rupee, or the proposed digital euro) will likely require coordinated international frameworks, potentially under institutions like the IMF, BIS, or G20. These developments could reshape geopolitical power dynamics and the role of traditional reserve currencies like the U.S. dollar.

For businesses, especially small and medium-sized enterprises, the adoption of QR code and digital payment systems presents both opportunities and growing pains. On the one hand, such systems can reduce overhead costs, improve transaction speed, and attract younger, tech-savvy customers. On the other hand, they often require upfront investment in software, hardware, employee training, and cybersecurity protocols—a burden that not all small vendors can easily afford.

The rise of digital payments also has implications for charitable giving, community fundraising, and informal economies. As physical cash becomes less common, cash-based donations or off-the-grid financial support may decline. Churches, non-profits, and grassroots causes will need to adapt by offering secure, user-friendly digital giving platforms, or risk alienating supporters who are moving away from physical currency.

These rapid changes raise a host of philosophical and spiritual questions. How will they impact our financial literacy and awareness of how money functions? Will they alter our psychological relationship to value, ownership, and generosity? Digital payments may render money "invisible" and abstract, potentially disconnecting people from mindful spending or budgeting practices.

In conclusion, the global adoption of QR codes and digital payments represents a watershed moment in the history of human commerce. While these technologies offer significant benefits—efficiency, accessibility, and innovation—they also bring profound challenges in terms of privacy, equity, cybersecurity, and spiritual discernment. As we move into this new financial frontier, it will be imperative for individuals, families, communities, and policymakers to proceed with wisdom, humility, and a commitment to ensuring that technological progress serves the best interests of all people—not just the powerful.

Let us remember the words of Proverbs 4:7:

"Wisdom is the principal thing; therefore get wisdom: and with all thy getting get understanding." In doing so, we can navigate these shifts faithfully—eyes open, hearts anchored, and souls aligned with eternal truths.

Chapter 35
The Teachings of Jesus Christ

Jesus' teachings form the cornerstone of the Christian faith and provide timeless guidance for navigating life's challenges—especially the turbulent times prophesied in Scripture concerning the end days. His words offer not only wisdom and moral clarity, but also comfort and direction for believers facing an uncertain future.

Jesus imparted crucial wisdom about life's priorities and eternity. He asked, in Matthew 16:26, KJV, "For what is a man profited, if he shall gain the whole world, and lose his own soul?" He also said, "For where your treasure is, there will your heart be also" (Matthew 6:21). These statements encourage deep reflection on what truly matters in life, urging followers to prioritize spiritual integrity and eternal values over temporary material gain.

Regarding His divine authority and the mission entrusted to His followers, Jesus declared in what is known as the Great Commission (Matthew 28:18–20): "All power is given unto me in heaven and in earth. Go ye therefore, and teach all nations, baptizing them in the name of the Father, and of the Son, and of the Holy Ghost: Teaching them to observe all things whatsoever I have commanded you: and, lo, I am with you always, even unto the end of the world." This commission emphasizes both the global scope of the Christian message and the comforting promise of Christ's abiding presence.

Jesus also warned of widespread deception (Matthew 24:4–5): "Take heed that no man deceive you. For many shall come in my name, saying, I am Christ; and shall deceive many". This warning is particularly relevant in an age of mass communication, digital misinformation, and spiritual confusion. Discernment, grounded in Biblical truth, is essential for navigating such times.

He foretold signs of the end, including wars, natural disasters, and betrayal among people (Matthew 24:6–8): "And ye shall hear of wars and rumours of wars: see that ye be not troubled: for all these things must come to pass, but the end is not yet... and there shall be famines, and pestilences, and earthquakes, in divers places. All these are the beginning of sorrows". He continued, "Then shall many be offended, and shall betray one another, and shall hate one another" (Matthew 24:10).

Jesus also predicted the rise of false prophets and a moral collapse:
"And many false prophets shall rise, and shall deceive many. And because iniquity shall abound, the love of many shall wax cold" (Matthew 24:11–12). Yet, despite these grim developments, He offered hope (Matthew 24:13): "But he that shall endure unto the end, the same shall be saved." This endurance is not passive but active—rooted in faith, love, and obedience to God's Word.

Speaking of the Great Tribulation, Jesus said (Matthew 24:21–22): "For then shall be great tribulation, such as was not since the beginning of the world to this time, no, nor ever shall be. And except those days should be shortened, there should no flesh be saved: but for the elect's sake those days shall be shortened." This passage underscores both the severity of the end times and God's mercy in limiting the duration for the sake of the faithful.

Vigilance is a central theme in Jesus' teaching (Matthew 24:42–44): "Watch therefore: for ye know not what hour your Lord doth come... Therefore be ye also ready: for in such an hour as ye think not the Son of man cometh". Just as a watchman stays alert to guard against a thief, so believers must remain spiritually prepared for Christ's return.

These teachings provide a comprehensive framework for end-time readiness. They emphasize spiritual preparedness, discernment, and unwavering faith. Jesus' words remind believers that while difficult times may arise, ultimate victory and eternal security belong to those who remain faithful to Him.

Moreover, Jesus' command to love remains central—even in tribulation (John 13:34). He said, "A new commandment I give unto you, That ye love one another; as I have loved you." This radical love includes forgiveness, grace, and even loving one's enemies (Matthew 5:44). In a world increasingly marked by division and hostility, such Christlike love becomes a powerful witness.

Jesus also clarified the nature of His kingdom (John 18:36): "My kingdom is not of this world". This truth reminds Christians that their primary allegiance is to God's eternal reign—not to earthly governments, ideologies, or systems of control. As global power structures become more centralized and potentially coercive, this teaching takes on renewed relevance.

In the Sermon on the Mount, Jesus laid out principles for righteous living: integrity, humility, mercy, peacemaking, and trusting God's provision (see Matthew 5–7). These teachings form a moral compass for navigating a corrupt world, offering peace and spiritual stability even amid global unrest.

Chapter 35
The Teachings of Jesus Christ

Many of Jesus' parables carried eschatological significance. In the Parable of the Ten Virgins (Matthew 25:1–13), He emphasized the importance of being spiritually prepared. In the Parable of the Talents (Matthew 25:14–30), He urged faithful stewardship of time, resources, and spiritual gifts—especially in times of uncertainty and testing.

Throughout His ministry, Jesus spoke of judgment and accountability. He warned of a day when the Son of Man would return to separate the righteous from the unrighteous, as illustrated in the Parable of the Sheep and the Goats (Matthew 25:31–46). These teachings underscore the eternal consequences of earthly choices and the need for continual repentance and obedience.

Finally, Jesus gave believers a model for how to pray:

"Thy kingdom come, Thy will be done in earth, as it is in heaven" (Matthew 6:10, KJV). The Lord's Prayer aligns the believer's desires with God's redemptive plan for humanity, acknowledging both His sovereignty and our dependence on Him.

In all of these teachings, Jesus provides enduring hope. While He doesn't shy away from describing the hardships and tribulations that will precede His return, He consistently reassures His followers of God's abiding love, sovereign power, and ultimate victory over evil. His promise in John 16:33 — "In this world ye shall have tribulation: but be of good cheer; I have overcome the world" —offers profound comfort and encouragement to believers as they navigate the uncertainties of the end times.

Jesus' teachings also emphasize the centrality of faith. He often praised individuals who demonstrated unwavering trust in God—such as the centurion who believed Jesus could heal from a distance (Matthew 8:10) or the woman with the issue of blood (Luke 8:48). In the context of end-time prophecies, this call to bold, persevering faith becomes even more vital. Believers are urged to trust God's plan and stand firm on His promises, even when chaos, persecution, or global upheaval tempt them toward fear or compromise.

The concept of forgiveness is another foundational theme in Jesus' ministry. He instructed His followers to forgive "seventy times seven" times (Matthew 18:22), signifying boundless grace. This radical forgiveness is especially relevant during times of tribulation, when betrayal, conflict, and moral collapse will tempt people toward bitterness and revenge. Practicing Christ-like forgiveness will be a powerful witness in an age of offense.

Jesus also promised the coming of the Holy Spirit, who would guide, comfort, and empower believers after His ascension. In John 14:26, He assured them: "But the Comforter, which is the Holy Ghost... shall teach you all things, and bring all things to your remembrance, whatsoever I have said unto you." This divine presence equips believers with wisdom and spiritual strength, especially during the trials of the end times.

In His teachings about the end of the age, Jesus stressed the suddenness and unpredictability of His return. In Matthew 24:37–39, He compared it to the days of Noah, when people were "eating and drinking, marrying and giving in marriage" until the flood came and took them all away. This comparison underscores the need for constant spiritual vigilance and moral readiness.

Jesus' teachings on stewardship and eternal priorities also carry weight for the end times. Matthew 6:19–21 says, "Lay not up for yourselves treasures upon earth... But lay up for yourselves treasures in heaven... For where your treasure is, there will your heart be also." Wise use of time, resources, and influence becomes even more critical as the world moves toward instability and spiritual deception.

Jesus displayed profound compassion for the marginalized and oppressed—whether healing lepers (Luke 17:12–19), defending women (John 8:1–11), or dining with sinners (Luke 5:29–32). His actions challenged societal and religious norms, offering a model of inclusive love and justice. In times of fear and division, this radical compassion must remain a guiding principle for believers, who are called to reflect His love even in a hostile world.

Jesus also taught candidly about the cost of discipleship. In Luke 9:23–24, He said, "If any man will come after me, let him deny himself, and take up his cross daily, and follow me." He warned that following Him would not always lead to worldly comfort or acceptance, but would often bring reproach, sacrifice, and even persecution. This sobering truth prepares the faithful to endure with courage and conviction when opposition arises.

In conclusion, Jesus' teachings offer a complete and timeless blueprint for living faithfully—not just in ordinary times, but also especially during the prophesied end times. They offer wisdom for navigating deception, encouragement for enduring trials, and hope for final redemption. As the world grows darker and more uncertain, His words remain an unshakable foundation for those seeking clarity, strength, and eternal purpose.

Chapter 36
The Four Horsemen and the Beasts of Revelation

The seven seals in the Book of Revelation represent a prophetic sequence of events unfolding on earth during the end times. Revelation 6:1–8 introduces the first four seals, describing the Four Horsemen of the Apocalypse, each symbolizing a distinct form of tribulation. These figures have long captivated the imagination of scholars and believers, offering both symbolic and potentially literal interpretations of the trials preceding Christ's return.

The First Horseman – White Horse (Conquest)

Revelation 6:2 says: "And I saw, and behold a white horse: and he that sat on him had a bow; and a crown was given unto him: and he went forth conquering, and to conquer." This rider has often been debated. While some view this figure as representing the Antichrist or deceptive conquest, others, as noted in this interpretation, see it as symbolic of imperial expansion, especially that of the British Empire and other colonial powers.

Historical records estimate that British colonial policies contributed to over 100 million deaths in India between 1881 and 1920, primarily through enforced famines, economic exploitation, and institutional neglect. The crown and bow could symbolize a form of conquest masked as civilization or order—dominance without immediate destruction.

Chapter 36
The Four Horsemen and the
Beasts of Revelation

The Second Horseman – Red Horse (War)

Revelation 6:4 states: "And there went out another horse that was red: and power was given to him that sat thereon to take peace from the earth, and that they should kill one another: and there was given unto him a great sword." This horseman is widely recognized as symbolizing global warfare.

The 20th century alone saw two world wars that killed over 100 million people, along with prolonged conflicts in Vietnam, Korea, Iraq, and Afghanistan. Many of these wars were facilitated or sustained through global financial systems, including those influenced by central banks such as the U.S. Federal Reserve, founded in 1913. The "great sword" represents both the physical violence and political powers that instigate widespread bloodshed.

The Third Horseman – Black Horse (Famine and Inequality)

Revelation 6:5–6 describes: "And I beheld, and lo a black horse; and he that sat on him had a pair of balances in his hand... A measure of wheat for a penny, and three measures of barley for a penny." The balances symbolize economic disparity and scarcity.

This rider is linked to major famines in the 20th century, such as the Great Leap Forward in China (1958–1962), which led to an estimated 30–45 million deaths, the Holodomor in Ukraine, and Bengal famine during WWII, each a product of economic mismanagement, ideological extremism, or colonial neglect.

The 2007–2009 Global Financial Crisis, caused largely by reckless banking practices in the U.S., affected billions worldwide, exacerbating poverty and inequality. The symbolism of this rider can extend beyond famine to include economic oppression and the global struggle for equity and justice.

The Fourth Horseman – Pale (or Green) Horse (Death)

Revelation 6:8 says: "And I looked, and behold a pale horse: and his name that sat on him was Death, and Hell followed with him..." The Greek word for "pale" is chloros, which can also be translated as green, symbolizing decay or disease. This rider brings mass death through plague, pandemic, and systemic collapse.

Pandemics like the Spanish Flu (1918, ~50 million deaths), HIV/AIDS (over 36 million deaths), and COVID-19 (which caused over 7 million confirmed deaths globally as of 2024) illustrate this prophecy's modern resonance. If geopolitical tensions in places like Ukraine and the Middle East escalate into global warfare, the potential for World War III raises the terrifying prospect of billions of casualties through nuclear or biological conflict.

The Beast from the Sea – Revelation 13:1–2

Beyond the seals, Revelation 13:1–2 introduces a powerful and ominous symbol: a beast with seven heads and ten horns, rising from the sea. It reads: "And I stood upon the sand of the sea, and saw a beast rise up out of the sea, having seven heads and ten horns, and upon his horns ten crowns, and upon his heads the name of blasphemy. And the beast which I saw was like unto a leopard, and his feet were as the feet of a bear, and his mouth as the mouth of a lion: and the dragon gave him his power, and his seat, and great authority."

Some scholars associate this beast (7 heads, 10 horns, and 10 crowns), 27 nations with a revived global political system, as possibly the European Union (EU).

The EU currently consists of 27 member nations. Notably, 10 countries. Czechia, Estonia, Cyprus, Latvia, Lithuania, Hungary, Malta, Poland, Slovenia, and Slovakia, joined simultaneously in 2004, aligning with the "ten kings" in Revelation 17:12, who "receive power... one hour with the beast."

The Vatican and the Healing of the Wound

Revelation 13:3 says: "And I saw one of his heads as it were wounded to death; and his deadly wound was healed." Some interpretations connect this with the Papacy, particularly the events of 1798, when Napoleon's general captured Pope Pius VI, temporarily stripping the Church of political power.

This "wound" was arguably "healed" with the 1929 Lateran Treaty, which reestablished Vatican City as a sovereign state under Mussolini's regime. Further prophetic resonance is seen in 1984, when the U.S. and the Holy See formally established diplomatic ties under President Ronald Reagan and Pope John Paul II, restoring Vatican influence on a global scale.

Today, the Holy See exercises global influence as both a religious and sovereign entity, participating in international diplomacy and intergovernmental organizations. Its unique dual status—governing both a religion and a state—sets it apart and aligns with Revelation's depiction of the beast with blasphemous power and broad authority.

Economic Control and the Power to Buy and Sell

Revelation 13:16–17 warns that the second beast, the United States, will eventually enforce a system in which "no man might buy or sell, save he that had the mark."

Modern developments in digital currency, Central Bank Digital Currencies (CBDCs), and cashless payment technologies are being scrutinized through this prophetic lens.

The global rise of QR code payments, biometric authentication, and digital ID systems create a potential infrastructure for economic exclusion—fulfilling the prophecy that economic participation could one day be conditional.

Diverse Interpretations and Ongoing Relevance

It's important to acknowledge that interpretations of Revelation vary widely. While some see direct historical and institutional correlations, others view the prophecies as symbolic, spiritual, or yet to be fulfilled in a future global system not yet fully realized.

Whether read as historical allegory, spiritual metaphor, or future roadmap, the messages in Revelation compel believers to remain watchful, discerning, and rooted in Christ.

The imagery of the beasts in Revelation draws rich parallels with the visions recorded in the Book of Daniel, forming a composite tapestry of prophetic symbolism that spans both the Old and New Testaments. This interconnectedness underscores the unity of biblical prophecy and reinforces the consistent themes of God's sovereignty, the rise and fall of earthly empires, and His ultimate victory over evil.

As we consider these prophecies, it's important to remember their central purpose: not to incite fear, but to offer hope, strength, and encouragement to believers facing persecution or uncertainty. Revelation was written to a church under duress, and its message remains timeless: God is in control, Christ has triumphed, and justice will prevail. While interpretations of how these prophecies will unfold may differ, the unshakable foundation of Christian faith is that God will overcome the forces of darkness and restore righteousness.

The beasts of Revelation—fearsome in appearance and commanding in influence—serve as stark warnings about the corrupting nature of unchecked power and the dangers of systems that rebel against God's authority. These beasts personify institutions, ideologies, and coalitions that seek to dominate politically, economically, and spiritually, often presenting themselves as saviors while working in opposition to divine truth. They remind believers to remain spiritually discerning and not place ultimate trust in human systems, political alliances, or charismatic leaders, but in the sovereignty of God alone.

At the same time, these prophecies demand wisdom and interpretive humility. Throughout church history, well-intentioned believers have sometimes misread or misapplied apocalyptic texts, leading to unnecessary panic, divisiveness, or even harmful actions. It is vital to approach these prophecies with a heart that is anchored in Scripture, guided by the Holy Spirit, and centered on the hope of redemption rather than the fear of destruction.

The symbolism of the Four Horsemen and the beasts also invites deeper reflection on the spiritual dimensions of evil and its many manifestations throughout human history. While these figures may correlate with specific empires or historical events, they also represent enduring archetypes: conquest, war, famine, death, and systemic rebellion. These are not merely historical realities, but ongoing spiritual forces—consequences of sin that continue to afflict humanity.

In this light, the Four Horsemen can be understood as symbolic of the global suffering that arises from human rebellion and brokenness. Their emergence signals the intensification of spiritual warfare and the culmination of history's long battle between the kingdom of God and the kingdom of darkness. This perspective calls believers to active engagement in spiritual warfare—through prayer, discipleship, sacrificial love, and resistance to evil in every form.

The beast rising from the sea, with its seven heads and ten horns (Revelation 13), represents a culmination of oppressive worldly systems—political, economic, and religious powers that align in opposition to God's kingdom. This beast is not merely a figure of brute force; it is a sophisticated, seductive power structure that demands allegiance and wages war against the saints (Revelation 13:7).

Its authority, derived from the dragon (commonly interpreted as Satan), emphasizes the spiritual undercurrent behind earthly events, showing that the ultimate battle is not merely political but cosmic in scale.

The prophecy about the beast's "deadly wound" being healed—and the world's amazement and worship that follow—serves as a solemn warning about the allure of evil disguised as revival or peace. Evil may appear defeated, only to rise again in a more powerful or deceptive form. This scenario reminds believers that external appearances can be misleading and that discernment is required to recognize spiritual deception. Even when evil seems to prevail, the faithful are called to endure, resist, and remain rooted in truth.

The mark of the beast, also described in Revelation 13, is closely linked to these prophetic symbols. While its exact form is still debated—whether literal, symbolic, digital, or biometric—it represents a forced alignment with a system that denies God's authority. The mark is ultimately about allegiance. To accept the mark is to willfully participate in a system that exalts itself above God and suppresses His people. The gravity of this decision is underscored by Revelation's severe warnings about its eternal consequences.

In contrast to the beast's system of coercion and fear, God's people are marked with His seal—a sign of ownership, protection, and divine purpose (Revelation 7:3; 14:1). This spiritual seal reflects the believer's identity in Christ and serves as a counter-testimony to the world's systems of control.

Conclusion

The Four Horsemen and the beasts of Revelation form a dramatic and sobering depiction of the final stages of history—a time when spiritual deception, suffering, and opposition to God will reach their peak. Yet they are not merely about catastrophe—they are also about the clarity that comes in contrast. Evil will be fully revealed, but so too will the faithfulness of God's people and the certainty of Christ's return.

Though their specific identities and timing are debated, the message is clear: Remain faithful. Remain alert. Do not be deceived. These prophecies are a divine call to resilient discipleship, reminding the Church of its role as a witness to truth in a world increasingly bent on falsehood.

In the end, Revelation is not a book of despair—it is a book of victory, worship, and hope. It closes not with destruction but with renewal: a new heaven and a new earth, where God dwells with His people and wipes away every tear. That is the promise believers hold fast to—even as the storm clouds gather.

Let's be clear that the first beast of Revelation is EU and the Vatican. The second beast is the United States. NATO is mostly an alliance between the United States and EU.

Chapter 37
The Reign of Seven Kings and the Future of Faith

Revelation 17 describes the reign of seven kings, which can be interpreted as a succession of empires that have shaped world history. This prophetic timeline provides a framework for understanding the progression of global powers leading up to the end times. Let's explore this succession of empires and their potential significance:

1. Egypt (Ramses 1279-1213 BCE): The ancient Egyptian empire, known for its advanced civilization and monumental architecture, represents the beginning of this prophetic timeline.
2. Assyria (Ashurbanipal 669-631 BCE): With its military might and cultural influence, the Assyrian Empire succeeded Egypt as a dominant power in the ancient Near East.
3. Babylon (Nebuchadnezzar the Great 605-562 BCE): The Babylonian Empire, famous for its hanging gardens and the exile of the Jewish people, followed Assyria.
4. Mede-Persia (Cyrus the Great? -530 BCE): This vast empire, which allowed the Jews to return from exile and rebuild Jerusalem, came next in succession.
5. Greece (Alexander the Great 336 – 323 BCE): The Greek Empire, spreading Hellenistic culture across a vast territory, followed the Mede-Persian Empire.
6. Rome (Augustus 31 BCE – 14 AD): The Roman Empire, which was in power during the time of Christ and the writing of the New Testament, succeeded Greece.

7. Ottoman Empire (36 Sultans between 1299-1922): Some interpretations suggest that the Ottoman Empire, which controlled much of the Middle East and southeastern Europe for centuries, represents the seventh king.

This progression of empires provides a historical context for understanding biblical prophecy and the development of world powers. While interpretations of the "seven kings" or empires described in the Book of Revelation may differ, many scholars agree that successive world empires have played defining roles in shaping prophetic history, especially those that oppressed God's people or opposed divine authority.

Turkiye, the modern successor to the Ottoman Empire, holds a unique and often overlooked position in both ancient and prophetic history. As the custodian of land that hosted some of the earliest known human settlements, Turkiye is home to Göbekli Tepe, a Neolithic site dated around 8000 BCE. Its massive stone structures and enigmatic carvings challenge conventional timelines and raise profound questions about early human spirituality and civilization.

Turkiye is also a land where Christianity and Islam intersect, a nation with deep Christian roots that now identifies primarily as an Islamic republic. This spiritual tension adds to its prophetic significance.

The region's connection to early Christianity is especially notable. The Council of Nicaea, convened in 325 AD under Emperor Constantine, was held in what is now northwestern Turkiye. This council was foundational in shaping orthodox Christian doctrine. Additionally, the island of Patmos, where the Apostle John received the visions recorded in Revelation, lies just off Turkiye's Aegean coast. Most significantly, the seven churches addressed in Revelation (Ephesus, Smyrna, Pergamum, Thyatira, Sardis, Philadelphia, and Laodicea) were all located in ancient Asia Minor—modern-day Turkiye. This geographic detail grounds key portions of Revelation in a very real and enduring region, suggesting Turkiye may yet play a pivotal role in end-time developments.

Turkiye's historical role as the seat of two influential empires—the Eastern Roman (Byzantine) Empire and the Ottoman Empire—continues to shape its identity. The Byzantine Empire preserved Christianity in the East for over a millennium, while the Ottoman Empire, following its conquest of Constantinople in 1453, expanded Islamic rule across three continents, enduring until 1923. This imperial legacy contributes to modern Turkiye's sense of destiny and geopolitical assertiveness.

In today's world, Turkiye stands at a strategic crossroads— literally and figuratively between East and West. It has been a member of NATO since February 18, 1952, traditionally aligning with Western powers. However, recent actions under its current leadership suggest a recalibration. Turkiye has expressed interest in joining BRICS, a growing economic coalition led by Brazil, Russia, India, China, and South Africa. BRICS extended invitations in 2024 to include Egypt, Iran, Saudi Arabia, Ethiopia, the UAE, and Argentina. This possible realignment could drastically reshape regional alliances and global power dynamics.

At the same time, Turkiye has spent over two decades attempting to join the European Union, with progress continually stalled. As the country with the second-largest military in NATO, Turkiye's shifting loyalties and potential dual alignments—with both BRICS and Western institutions—could signal the emergence of a new global coalition or transitional phase.

This brings us back to the mysterious "woman in purple and scarlet" who rides the beast in Revelation 17—a symbol widely interpreted as a powerful religious-political system that dominates and influences kings and nations. Revelation 17:11 states:

"And the beast that was, and is not, even he is the eighth, and is of the seven, and goeth into perdition."

This enigmatic passage suggests the rise of an eighth kingdom, emerging from one of the previous seven, yet distinct in its nature and ultimate fate. While this eighth system is not fully manifest, we may be witnessing its formation in transitional alliances, centralized global authority, or unprecedented technological control.

Revelation continues:

"These have one mind, and shall give their power and strength unto the beast. These shall make war with the Lamb, and the Lamb shall overcome them: for He is Lord of lords, and King of kings: and they that are with Him are called, and chosen, and faithful." (Revelation 17:13–14)

This prophecy reveals a final coalition of powers that will unite in opposition to Christ and His followers. Though they appear invincible, they are destined for defeat. The Lamb—Jesus Christ—will overcome them, and with Him will be the faithful remnants, who did not yield their allegiance to the beast.

Chapter 37
The Reign of Seven Kings and the
Future of Faith

Conclusion

Turkiye's spiritual and historical legacy, its imperial past, and its modern geopolitical evolution all contribute to its potential significance in end-time prophecy. Its proximity to ancient Christian sites, its pivotal role in early church history, and its central place in current global affairs may signal that it is more than a bystander in the unfolding drama of Revelation.

In 2025, Turkiye is host to peace negotiations between Russia and Ukraine to end the war that already killed or injured more than a million people.

Whether Turkiye emerges as a central player in the eighth kingdom or remains a bridge between power blocs, its story reminds us that biblical prophecy is not abstract—it is rooted in real lands, real leaders, and real decisions. And above all, the battle lines drawn in Revelation are not merely geopolitical—they are spiritual, calling every believer to stand firm, remain faithful, and keep their eyes fixed on the Lamb who will ultimately reign over all.

Chapter 38
The Catholic Church and Pedophilia

Revelation 17:3–4 declares:

"So he carried me away in the spirit into the wilderness: and I saw a woman sit upon a scarlet coloured beast, full of names of blasphemy, having seven heads and ten horns. And the woman was arrayed in purple and scarlet color, and decked with gold and precious stones and pearls, having a golden cup in her hand full of abominations and filthiness of her fornication."

These verses offer some of the most vivid and controversial imagery in the Book of Revelation, often interpreted as a symbol of a powerful religious system intertwined with worldly political powers. The purple and scarlet attire is especially significant—colors traditionally associated with authority, wealth, and religious ceremony.

In the Roman Catholic Church, bishops wear purple, while cardinals wear scarlet red. This combination is unique among religious institutions and aligns strikingly with the description in Revelation. Additionally, the adornment with gold, precious stones, and pearls mirrors the opulence and pageantry often observed in ecclesiastical traditions, particularly in the Vatican.

The City on Seven Hills

Revelation 17:9 adds another clue:

"And here is the mind which hath wisdom. The seven heads are seven mountains, on which the woman sitteth."

281

Many scholars and commentators throughout church history have connected this verse to Rome, which historically was built upon seven hills: Palatine, Aventine, Caelian, Esquiline, Viminal, Quirinal, and Capitoline. This geographical detail is often cited as evidence that the prophetic "woman" represents a religious-political power headquartered in Rome.

Revelation 17:18 strengthens this association:

"And the woman which thou sawest is that great city, which reigneth over the kings of the earth."

The only city in the world that functions as both a religious headquarters and an independent sovereign state is Vatican City, the administrative and spiritual center of the Roman Catholic Church. It maintains diplomatic relations with over 180 countries and holds observer status in the United Nations.

Systemic Abuse and the Apostate Church

Despite its spiritual claims, the Catholic Church has been plagued by decades of sexual abuse scandals, involving thousands of priests and devastating hundreds of thousands of victims.

- On February 28, 2024, a search for "How many pedophile priests have gone to jail since Pope Francis took over the Vatican?" revealed virtually none.
- On October 5, 2021, Al Jazeera reported that an investigation in France revealed more than 216,000 victims of sexual abuse by around 3,000 clergy and church workers between 1950 and 2020. According

to Jean-Marc Sauvé, the abuse was "systemic"—and 80% of the victims were boys.

- On October 27, 2023, the BBC published a damning report: more than 200,000 children were sexually abused by Spain's Catholic clergy. Spain's ombudsman condemned the Church's culture of denial and cover-up, calling it a "devastating" breach of trust.
- In the United States, Catholic dioceses have paid over $3 billion in settlements. The infamous Boston Globe investigation (2002), later dramatized in the film Spotlight, uncovered widespread abuse and systematic concealment by Church leadership.

In 2001, Pope John Paul II apologized for abuse within the Church, calling it a "profound contradiction" of Christ's teachings. But his critics argue he did not act swiftly or decisively to protect victims. His offer of indulgences in 2000, just before the flood of lawsuits and settlements began, raises troubling questions about the Church's awareness of the scandal's scope.

Pope Benedict XVI expressed "shame" over the crisis and met with some victims, but under his leadership, the Church's accountability mechanisms remained opaque.

When Pope Francis was elected in 2013, hopes ran high that he would finally implement reform. But when asked shortly after his election whether the new Pope would truly "clean up the mess," a Jesuit priest at Santa Clara University responded simply: "Nothing will happen."

Indeed, the problem persists. In 2024, survivors and advocates continue to protest the lack of criminal accountability and the Church's slow progress on transparency.

Apostasy and False Teaching

Pope Francis has also drawn criticism for interfaith statements that appear to contradict foundational Christian doctrine. In a September 2024 visit to Singapore, a first papal visit in nearly 40 years, he told a group of young people:

"All religions are paths to God... There is only one God, and religions are like languages, paths to reach God."

This statement directly contradicts Jesus' words in John 14:6:

"Jesus saith unto him, I am the way, the truth, and the life: no man cometh unto the Father, but by me."

Such ecumenical pluralism aligns with the "apostate church" described in Scripture—a church that claims religious authority but denies essential truths. In 2 Thessalonians 2:3, Paul warned that a great falling away—apostasia—would precede the rise of the antichrist. Apostasia is a Greek word that translates to apostasy, meaning the abandonment of a religious or political belief, or a formal renunciation of a faith.

The Catholic Church is not alone in this drift. Many Protestant denominations have also embraced doctrines and ideologies contrary to Scripture. The apostate church, therefore, is not defined by denomination but by its rejection of Biblical truth, even while maintaining a religious appearance.

Revival in the Global South

Amid this darkness, the light of the Gospel continues to spread. Christianity is growing rapidly in regions often overlooked by Western institutions:

- In **Africa**, the number of Christians has risen from about **10 million in 1900 to 734 million in 2024**.
- In **Latin America**, millions remain faithful to Christian doctrine despite cultural and institutional challenges.
- Even in **China**, where Christianity is heavily persecuted, **underground house churches** are thriving.

According to the PEW Research Center, "As of 2010, about a quarter of the global Christian population was in Europe (26%), a quarter in Latin America and the Caribbean (25%) and a quarter in sub-Saharan Africa (24%). Significant numbers of Christians also live in Asia and the Pacific (13%) and North America (12%). Less than 1% live in the Middle East-North Africa region, where Christianity began. Sub-Saharan Africa is predicted to have the largest share of the world's Christians, rising from 24% in 2010 to 38% in 2050, Europe's share is projected to drop to about 16% from 26%. Latin America and North America are also expected to see modest declines in their respective shares of the global Christian population.

Conclusion

The prophetic image of the woman clothed in scarlet and purple, seated on a beast with seven heads and ten horns, is one of seduction, corruption, and spiritual deception. Many elements—color symbolism, geographic location, global influence, religious opulence, and systemic abuse—appear to converge in the modern Roman Catholic Church, though this interpretation is not without controversy.

Yet, the warning is not limited to any one institution. The real danger lies in any religious system that rejects the authority of Christ, prioritizes power over truth, and covers sin rather than repenting of it.

Still, the Gospel continues to advance. And as Revelation promises, the Lamb—Jesus Christ—will overcome every false power. His followers, the "called, and chosen, and faithful," will stand victorious with Him.

"Come out of her, my people, that ye be not partakers of her sins." (Revelation 18:4)

Chapter 39
Evidence of End-Time Prophecies

Isaac Newton, one of the most outstanding scientists, needs no introduction. Here, we'll focus not on all the remarkable achievements he had in physics and mathematics but on his research rooted in a personal pursuit of God. In fact, Newton wrote more than a million words on theology—far more than he ever published on science. He devoted more time to biblical exegesis and alchemical studies than to physics and mathematics.

In particular, he spent most of his time studying the Book of Daniel and the Book of Revelation in the Bible. These two books are about end-time prophecies, and I'll share with you some of the prophecies that have been fulfilled. I'll also discuss what is being or about to be fulfilled. Newton saw these prophetic texts not merely as spiritual metaphors, but as cryptic records of real historical processes and future world events. He approached them with the same logical rigor he applied to science.

Newton has been the most influential scientist of all time, formulating the laws of mechanics, the law of universal gravity, and other laws to describe everyday objects in motion to rockets taking off for space exploration. Without Isaac Newton's insights, we could not have the Hubble Telescope or the James Webb Space Telescope (JWST). His invention of the reflecting telescope in 1668 eliminated chromatic aberration found in refracting telescopes and became the foundation for modern astronomical instruments.

I would not have seen the Andromeda Galaxy without looking through a 22" Reflector Telescope if not for his reflector telescope invention. The Andromeda Galaxy, our nearest galactic neighbor, is located about 2.5 million light-years away and can be observed using large reflecting telescopes—technology made possible by Newton's design innovations.

He also co-invented calculus, without which there would be no physics or engineering. Although the invention of calculus was contested by Gottfried Wilhelm Leibniz, both Newton and Leibniz independently developed the framework that underpins modern mathematics, physics, and engineering.

Isaac Newton started his most prolific scientific work during the Great Plague of London, lasting from 1665 to 1666. The plague killed an estimated 100,000 people, almost a quarter of London's population, in 18 months. He spent two years at his mother's farm in Woolsthorpe, hiding from an outbreak of the plague that took hold of Cambridge, where he was a student. This period became known as Newton's "Annus Mirabilis" (year of wonders), during which he developed early ideas about gravity, optics, and the foundations of calculus.

To Isaac Newton, science was a portal to God's mind, a bridge between humans and the Divine. Newton, a name that represents the quintessential rationalist, was, in fact, a rational mystic. He believed that science was like a religious practice, a meeting with God's mind. In Newton's view, natural laws were not autonomous—they were decrees issued and upheld by God.

In studying the Book of Daniel and the Book of Revelation during the Great Plague, Newton surely wondered whether he was living in the end times. He calculated the year when the end would come. The most definitive date he set for the apocalypse was 2060. This calculation was based on his interpretation of Daniel 12:7 and Revelation 11:2–3, and his belief that the end times would follow 1,260 years after the establishment of the papacy—placing the culmination around the year 2060. Newton was careful to state that this was not a prediction of the end of the world, but the end of ecclesiastical corruption and the beginning of a renewed era.

On September 23, 2017, the Bible's Book of Revelation 12:1-2 was fulfilled in a spectacular event that could be seen by millions of people. On that date, many Christian prophecy watchers noted a rare astronomical alignment involving the constellation Virgo (the Virgin), the sun, the moon, and the planet Jupiter. They interpreted this as a fulfillment of Revelation 12:1–2, which describes a 'woman clothed with the sun, with the moon under her feet, and a crown of twelve stars on her head.'

The previous time such an event happened was seven thousand years ago. Astronomers clarified that while elements of this alignment have occurred before, the 2017 arrangement was especially unique, prompting much public attention. However, mainstream scientists and theologians remain divided on whether this event constituted an actual prophetic fulfillment.

The prophecy said, "And there appeared a great wonder in heaven; a woman (Constellation Virgo) clothed with the sun, and the moon under her feet, and upon her head a crown of twelve stars (nine stars of Constellation Leo with Mercury, Venus, and Mars). And she is being with child (Jupiter inside Virgo) cried, travailing in birth, and pained to be delivered." This exact cosmological alignment was a sign of the end time prophesied in the Bible about two thousand years ago.

Chapter 39
Evidence of End-Time Prophecies

A total solar eclipse is a rare event that happens once every 1–2 years somewhere on Earth. However, the path of totality—a narrow corridor where the total eclipse is visible—only covers a small portion of the planet each time, making any specific location's experience of a total eclipse extremely rare. Two total solar eclipses (seven years apart) that mark an X in the middle of a specific country that will play a key role in the end time is extremely rare.

The total solar eclipse of August 21, 2017, moving from northwest (Oregon) to southeast (South Carolina), and the total solar eclipse of April 8, 2024, spanning from southwest (Texas) to northeast (Vermont), mark an X in the middle of the United States. This intersection occurs near the town of Makanda in southern Illinois, close to the New Madrid Seismic Zone—an area known for catastrophic earthquakes in the early 1800s. In 2017, I drove from California to Yellowstone Park in Wyoming to watch the total solar eclipse, which blew my mind. When I saw the total eclipse, I experienced something like a flash (not exactly) in my mind. The most accurate way to describe that experience is that it literally blew my mind.

On April 8, 2024, millions of people watched the total eclipse in person in Mexico, the US, and Canada. Mazatlán, Mexico, was the first place it arrived at, and many people enjoyed it. The Indianapolis Motor Speedway, where the world-famous Indy 500 annual automobile race is held, and Niagara Falls were some of the most popular spots where people watched the total eclipse. NASA reported that the eclipse path covered over 15 U.S. states and was observed by more than 30 million people in person, with tens of millions more watching online. I liked the beaches in Mazatlán, the Indy 500 race, and Niagara Falls when I was there years ago.

One of the most important end-time prophecies soon to be fulfilled is written in the Book of Revelation 13:14–18: "And deceiveth them that dwell on the earth by means of those miracles which he had the power to do in the sight of the beast; saying to them that dwell on the earth, that they should make an image to the beast, which had the wound by a sword, and did live."

The United States is the Second Beast of Revelation. It is the beast coming up out of the earth, and he has two horns like a lamb, and he spake like a dragon. Coming up out of the earth means a place lightly populated when Revelation was written almost two thousand years ago. Jesus Christ was the Lamb of God. Christians founded the United States of America. The Continental Congress adopted the Declaration of Independence on July 4, 1776.

On June 21, 1788, the Constitution was ratified. In 1791, the First Amendment to the Constitution affirmed the separation of church and state. Unlike the first beast with ten crowns, the lamb has no crown. The US has no king. It is a republic. The two horns may symbolize its dual founding principles—civil liberty and religious freedom. In 1955, 92% of the US population was Christian. He spoke as a dragon, which is in line with being the only superpower in the world since the early 1990s. A multi-polar world, however, is emerging. Since the collapse of the Soviet Union in 1991, the United States has maintained global military and economic dominance, though the rise of China, Russia, and BRICS nations now signals a shift toward multipolarity.

The United States created the Federal Reserve, which facilitated WWI, WWII, and many other wars, killing over 100 million people. The Federal Reserve was created in 1913 as the central banking system of the United States.

Though not directly causing wars, its monetary policies have played roles in funding U.S. military engagements and shaping global economic trends. The US caused the "Great Depression" and "Great Recession."

The Great Depression began with the U.S. stock market crash in 1929, spreading globally. The 2008 financial crisis, triggered by risky mortgage-backed securities and lax regulation in the U.S. housing market, nearly collapsed the global financial system. The Federal Reserve and major U.S. banks were central actors in both crises.

The global financial system was only days away from a total meltdown in 2008 due to America's corruption. Former U.S. Treasury Secretary Henry Paulson and Federal Reserve Chair Ben Bernanke later admitted the global financial system was within hours of collapse during the Lehman Brothers failure. It is guilty of the seven deadly sins.

Greed, pride, envy, wrath, lust, gluttony, and sloth—often considered symbolic rather than literal indictments—are increasingly cited by moral commentators as descriptors of modern Western excess, particularly in critiques of American materialism, military expansion, and cultural exports.

Greed – The top 1% of Americans have 10 times more wealth than the bottom 50%.

Anger – highest homicide rate in the world

Lust – Adultery, fornication, and pornography are rampant

Gluttony – Eating and drinking excessively, with two-thirds of the people obese or overweight

Envy – People having a mostly negative feeling of desire for something that someone else has, and you do not

Sloth – Apathy, indifference, and depression that can lead to unfulfilled duties and obligations for many people

Pride – The strongest military superpower in the world and the country with the highest GDP, neglecting its $37 trillion debt.

The first Beast of Revelation is more complicated. There are 27 nations in the beast with seven heads, ten horns, and ten crowns. The only entity with 27 nations is the European Union. In Revelation 13:1–2, the beast is described as rising from the sea—often interpreted symbolically as peoples, nations, and languages. The seven heads and ten horns echo imagery in the Book of Daniel, linking ancient empires with modern systems of global governance. Vatican City, an independent country and the smallest nation in the world, is surrounded by Italy. Though not a member of the European Union, Vatican City maintains diplomatic relations with the EU and shares geographical, cultural, and economic ties with Europe.

The European Union Constitution doesn't even mention God, as if God doesn't exist. This omission was a deliberate decision made during the drafting of the Treaty establishing a Constitution for Europe in 2004, following contentious debates among member states about the role of Christianity in Europe's history. Instead, it refers only to the "cultural, religious and humanist inheritance of Europe."

If there is no God, there are no Ten Commandments. Does it make it easier for the EU to steal Russian government money from its banks? In 2022, following Russia's invasion of Ukraine, the EU froze over €300 billion of Russian Central Bank assets. As of 2024, the EU is considering using interest generated from these frozen assets to support Ukraine, which has sparked international legal debates.

Nowadays, worshippers of God rarely attend most of the great churches in Europe. Church attendance has declined across Western Europe, with Pew Research (2018) showing that only around 10–20% of Christians in countries like Germany, France, and the Netherlands attend church weekly.

Will the EU be better off under Darwinism or under God? While secularism and evolutionary science dominate public discourse in much of Europe, discussions about moral authority, tradition, and spiritual meaning continue in various academic and religious circles. In the US, within the last 25 years, the number of respondents who say that religion is "very important" to them has declined from 62 percent to 39 percent. (Source: Gallup, 2023.) It is hard to imagine that the EU has become a common thief.

The European Union has had a unique partnership with 27 European countries since 2013, known as Member States or EU countries. Together, they cover much of the European continent. The EU is home to around 447 million people, which is around 6% of the world's population. The UK left the EU in 2016, and Sweden joined the NATO military alliance on March 7, 2024, not the EU, as it has been a member of the European Union since 1995.

The EU countries are: Austria, Belgium, Bulgaria, Croatia, the Republic of Cyprus, Czech Republic, Denmark, Estonia, Finland, France, Germany, Greece, Hungary, Ireland, Italy, Latvia, Lithuania, Luxembourg, Malta,

Netherlands, Poland, Portugal, Romania, Slovakia, Slovenia, Spain, and Sweden.

Are the European Union and the Papacy parts of the first beast of Revelation? Many Bible prophecy scholars debate this. Some view the EU and Papacy as symbolic of revived Roman imperial and religious power. Revelation 17:9 mentions "seven mountains," which some interpret as the seven hills of Rome—possibly referencing the Vatican's influence. However, interpretations vary across denominations and theological traditions.

Since Donald J. Trump became US President for the second time on January 20, 2025, the world has been turned upside down. The United States withdrew from the World Health Organization and the Paris Climate Accords. He instigated a trade war with Canada and Mexico by proposing tariffs and installing new tariffs on China. He wants to annex Canada, Greenland, and the Panama Canal. He proposed forcibly relocating the Palestinian population from the Gaza Strip to other Arab states and rebuilding Gaza into a tourist destination. These policy developments, though hypothetical as of mid-2025, reflect an imagined second Trump term based on his previous foreign policy postures.

No official record exists of annexation plans for Canada or Greenland, though Trump expressed interest in purchasing Greenland during his first term.

The Trump administration suspended all military aid to Ukraine, offered concessions to Russia, and requested half of Ukraine's oil and minerals as payment for U.S. support. On February 28, 2025, he publicly had a blowup with Ukraine President Zelenskyy in the Oval Office in front of the press.

Trump accused Zelenskyy of potentially causing WWIII. He also established the Department of Government Efficiency, or DOGE, led by Elon Musk, to cut spending, limit federal bureaucracy, and oversee mass layoffs across federal agencies.

DOGE is a fictional agency; however, Elon Musk has expressed interest in government reform and efficiency. His administration has also publicly rebuked NATO and the EU. These are just some of the glaring changes in just six weeks. It is an understatement to say turbulent times are ahead for the world.

Europe's three most powerful nations, Germany, France, and the United Kingdom, are afraid that, without the US's full commitment to NATO, Russia will invade them in the future. Are they afraid because the armies of Hitler and Napoleon invaded Russia in the past, and Winston Churchill asked for American participation in invading Russia in 1945? There is no confirmed historical evidence that Churchill formally proposed invading the Soviet Union in 1945, but Operation Unthinkable was a British military contingency plan to confront the USSR, declassified in 1998.

In Germany's 2025 election, Friedrich Merz's conservatives won, but Alternative for Germany (AfD) doubled its support in just four years to 20.8% and became the second biggest political force in parliament. Outgoing Chancellor Olaf Scholz's SPD had its worst performance in decades, with only 16.4%. Under Merz's leadership, Germany's relationship with the US may not be smooth. Merz, as a CDU leader, advocates a strong transatlantic alliance, but internal EU divisions and populist pressures, such as the rise of AfD, complicate foreign policy consistency.

The EU is trying to create an army of its own, despite opposition from some of its members. The concept of a European Union defense force has been debated for decades. The EU's Permanent Structured Cooperation (PESCO), launched in 2017, is a step in that direction. However, NATO remains the dominant defense alliance, and several EU members—like Poland and the Baltic states—prefer continued reliance on NATO over a unified EU army.

We are living in challenging times. We must strive and pray for peace.

When you need God's protection, provision, and guidance, remember Psalm 23 (A Psalm of David).

"The LORD is my shepherd; I shall not want. He maketh me to lie down in green pastures: he leadeth me beside the still waters. He restoreth my soul: he leadeth me in the paths of righteousness for his name's sake. Yea, though I walk through the valley of the shadow of death, I will fear no evil: for thou art with me; thy rod and thy staff they comfort me. Thou preparest a table before me in the presence of mine enemies: thou anointest my head with oil; my cup runneth over. Surely goodness and mercy shall follow me all the days of my life: and I will dwell in the house of the LORD forever." (Psalm 23, KJV)

Your journey on this earth began as a vulnerable baby, and your life itself is a miracle. Be grateful to all who cared for you—parents, guardians, mentors, and even strangers who played a part in your survival and growth.

Globally, an estimated 73 million abortions occur each year, according to the World Health Organization (WHO). This number eclipses the death tolls from wars and pandemics combined. Congratulations to you—you are among those who were given the gift of life. You get to live.

You were fortunate to be born. You are alive, breathing, and hopefully healthy. That, in itself, is cause for gratitude and celebration. Every choice you make transforms both yourself and your surroundings. How you live defines and reveals the identity of your soul.

Everyone is entitled to believe what he or she wants. Respecting others' beliefs is part of living in harmony. But that does not mean we abandon judgment altogether. Sound, righteous judgment—rooted in wisdom—is vital. The Bible itself teaches that we are to judge rightly (John 7:24), though ultimate judgment belongs to God.

This book is, in essence, about discovering the truth—not just as an abstract concept, but also as a personal journey of identity and purpose. It is also about rediscovering why we were born—why we are here on this earth.

Some people don't believe they were born with a purpose. But those who know who they are often have a clear sense of why they are here. Identity and purpose are intertwined. If your identity has been defined by others—parents, peers, society, or institutions—then your purpose often feels uncertain, even irrelevant.

Many people live with confusion about their true nature. Lost in the mist of illusion and the seduction of the material world, they wander without spiritual direction. These are the souls who grieve not just their spirit, but also the Spirit that longs to guide them. They are lost in the forest of worldly desires on the island of death, unaware that just beyond the shadows lies the sea of eternity—an endless ocean of divine meaning, purpose, and peace.

For those deeply immersed in fantasy worlds—whether through video games, media, or addictive distractions—truth can seem distant or irrelevant. If this describes you, my prayer is that you find clarity, even if only in a single moment of silence. That spark can ignite a lifetime of change.

We can choose what to believe, but our choices do not alter the truth itself. Each of us sees reality through a particular lens. Some call it God. Others call it Nature. Some invoke the Universe, Consciousness, or even Chaos. But all of us, in our own way, are trying to describe the same mysterious Source behind existence.

I once visited a Mayan temple complex with a long stone path. As I walked, I noticed I couldn't see the path clearly ahead. The sun was behind me, and my own shadow fell directly in front of me—obscuring my view. It struck me then: I was the one blocking my path. How often do we do the same in life—letting our own fears, ego, and doubts obscure the way forward? Physically, emotionally, spiritually—what parts of you are casting shadows on your destiny?

Jesus said, "I am the way, the truth, and the life." What is blocking you from seeing Him for who He claimed to be? We often hear that there are many ways to do something—and yes, in tasks, that may be true. But in life's most essential matters—truth, salvation, peace—the path may not be as wide as we assume.

Climbers ascending Mount Everest from the Nepal side follow a narrow, treacherous route through the Hillary Step.

On the Tibet side, the route is different, equally dangerous, and also leads to the summit. The lesson? There may be multiple paths to the mountain, but each requires commitment, humility, and perseverance. Your spiritual path is no different. Choose wisely.

Each day, I practice spiritual surrender. I let go of everything—beliefs, relationships, material possessions, even my identity—if only for a moment. I release my anxieties, shame, pride, and every form of emotional weight. I release the seven deadly sins: pride, greed, wrath, lust, gluttony, envy, and sloth. In doing so, I clear the space in my soul to start again—with God as my foundation.

I reflect on the moments when I have truly felt God's presence—quiet, powerful, unmistakable—and also the moments when God seemed silent or absent. I carry both. They shape my faith. In prayer, I open my day. Not with demands, but with awareness. Not for blessings alone, but for the strength to walk in God's presence—step by step, moment by moment.

Chapter 40
Reflections on Major Transitions and the Future

Would Catholicism continue perpetually, or will "The Last Pope" exist in our lifetime? According to the 12th-century "Prophecy of the Popes" attributed to Saint Malachy, some believe Pope Francis was the final pontiff before the end times—though the Vatican has never endorsed this prophecy. Among the one billion Catholics, how many can name even twelve cardinals in the top hierarchy of the Vatican? There are currently over 200 cardinals globally, but only about 120 are eligible to vote in a papal conclave. The average lay Catholic remains largely unfamiliar with the College of Cardinals, reflecting the distance between hierarchy and laity.

Pope Francis, who changed the church forever, died on April 21, 2025, and many books will be written on what he did to change the world. He was known for championing environmental stewardship (as seen in Laudato Si'), interfaith dialogue, and progressive stances on social issues like LGBTQ+ inclusion and economic inequality. His efforts to decentralize church authority and push for synodality redefined papal leadership in the modern age.

When the enlightened Dalai Lama dies, who can adequately replace him as the spokesman for Buddhism? The current 14th Dalai Lama, Tenzin Gyatso, has suggested the possibility that his reincarnation may not continue or could be chosen outside Tibet—potentially even as a woman—challenging centuries-old tradition. If the previous Dalai Lamas keep reincarnating, how would their souls ever achieve Nirvana? In Tibetan Buddhism, the bodhisattva ideal explains this: enlightened beings voluntarily delay their own final liberation in order to serve others across lifetimes. This self-sacrificial cycle is not seen as failure to attain Nirvana, but as the highest spiritual service. The Dalai Lama is already in his nineties and many books will also be written about him. Born in 1935, his impact on global peace, Tibetan identity, and interreligious harmony will endure long after his passing.

When the Dalai Lama came to Santa Clara University in 2014 to speak on the topic of "Compassion, Business, and Ethics," I received an invitation to attend as an advisory board member of the distinguished Ignatian Center for Jesuit Education and I learned from the wisdom he shared.

The Nature of Life and Remembrance

Empires rise and fall. Many have lived, but few are remembered. We are cast members in an eternal play, seeing scenes and characters change, but the play remains. From Rome to the British Empire, history shows that no human institution is eternal. Yet ideas and legacies— spiritual, cultural, and moral—can outlast stone and steel.

The Challenges of the 21st Century

We live in an age when global disasters increase. The 21st century has been a turbulent one to date, and it could get worse. As people cope with, all at the same time, global heat waves, floods, drought, and wildfires, as well as financial pressures and health issues, things can get out of control very quickly in an age of instant news and reactions to that news.

The United Nations has declared climate change the defining issue of our time, citing that over 3.6 billion people live in areas highly vulnerable to environmental disasters. The World Health Organization has also noted that climate-related health threats are on the rise, including malnutrition, heatstroke, and infectious disease.

At a spiritual level, many Christians see our time as signs of the Second Coming of Christ as they witness many prophesies getting fulfilled according to the books such as Matthew, John, Daniel, and Revelation in the Bible. Events like wars, famines, and earthquakes—mentioned in Matthew 24—are often cited by eschatologists as signs, though such interpretations have existed across centuries.

Political institutions are also facing scrutiny and change. With Queen Elizabeth II's passing, will the monarchy survive in the United Kingdom? King Charles III ascended the throne in 2022, but support for the monarchy has declined among younger generations, particularly in Commonwealth countries like Jamaica, Australia, and Canada, where republican movements are gaining traction.

Will countries worldwide demand reparations from the Monarchy for profits gained from slavery and natural resources in Africa, the plundering of $450 trillion from India, and the opium wars that humiliated China for a century?

While the $450 trillion figure regarding India is debated, economic historians like Utsa Patnaik estimate Britain extracted roughly $45 trillion in wealth during colonial rule. Recent years have seen growing calls for reparations and formal apologies, especially from Caribbean nations and former African colonies.

These issues reflect ongoing debates about historical injustices and the role of traditional institutions in the modern world. These questions reflect a quest for truth in an era where omissions and lies often obscure reality.

The 21st century has indeed been marked by turbulence and rapid change. We've witnessed increased global disasters, from devastating natural events to worldwide health crises. The interconnectedness of our world means that local events can quickly have global repercussions, leading to financial pressures and societal upheavals that affect people across the planet. The COVID-19 pandemic starkly demonstrated this interconnectedness, triggering a global economic slowdown and transforming education, labor markets, and healthcare systems.

It's crucial to approach the challenges of the 21st century with wisdom and discernment. Throughout history, believers have seen signs of the End Times in their own era, and it's important to balance prophetic understanding with practical engagement in the world around us. Theologians like N.T. Wright and John Piper emphasize preparedness over speculation, urging Christians to live faithfully regardless of eschatological timelines.

As we consider the future of faith and the potential fulfillment of biblical prophecies, it's essential to remember that our primary calling as believers is to love God and love our neighbors. Regardless of how end-time events unfold, we are called to be lights in the darkness, sharing God's love and truth with those around us.

The rapid pace of technological advancement also raises questions about the nature of humanity and our relationship with God. As artificial intelligence and biotechnology progress, we may face new ethical dilemmas and challenges to traditional understandings of human nature and consciousness.

Bioethicists and theologians alike are now debating issues such as brain-computer interfaces, CRISPR gene editing, and AI-driven worship content. The Vatican's Pontifical Academy for Life has even issued guidelines on AI ethics. How will faith traditions adapt to these changes while maintaining their core principles?

Climate change and environmental degradation present another set of challenges that intersect with religious and ethical considerations. Many faith traditions emphasize stewardship of the Earth, but how will they respond to the urgent need for global action on climate issues? Pope Francis's encyclical Laudato Si' and initiatives like GreenFaith and Islamic Declaration on Climate Change signal a growing interfaith environmental movement. Will we see a rise in eco-theology or new religious movements centered on environmental concerns?

The increasing secularization of many societies, particularly in the West, raises questions about the future role of organized religion in public life. Will we see a continued decline in religious affiliation, or might there be a resurgence of faith in response to global challenges and uncertainties?

Surveys by Pew Research suggest that "nones" (religiously unaffiliated) are growing, especially in the U.S. and Europe, but global religiosity remains high, particularly in Africa, the Middle East, and Southeast Asia. Crises often prompt a spiritual revival, even in secular cultures.

At the same time, we're witnessing the rapid growth of Christianity in the Global South, particularly in Africa and parts of Asia. This shift in the center of gravity of the Christian world may lead to new theological emphases and cultural expressions of faith. By 2050, Africa is expected to be home to over 40% of the world's Christians, and Pentecostal movements are rapidly shaping doctrines, worship, and political engagement across the continent. How will this impact global Christianity and its relationship with other world religions?

Interfaith dialogue and cooperation may become increasingly important in a world facing global challenges that require collective action. Will we see more collaboration between different faith traditions, or will religious differences continue to be a source of conflict? Organizations like the Parliament of the World's Religions and the United Nations Alliance of Civilizations are fostering such cooperation, especially on issues like climate justice, refugee crises, and human rights.

The digital age has also transformed how people engage with faith and spirituality. Online communities, virtual church services, and religious apps have become increasingly common. How will these technological changes shape religious practice and community in the coming decades?

Digital platforms have enabled hybrid forms of worship, real-time Bible study, and global outreach, but they also pose risks of disconnection, disinformation, and consumeristic spirituality. Future generations may redefine "sacred space" entirely.

The war in Ukraine has severely impacted Germany's industrial base, which had long relied on cheap Russian gas for energy-intensive sectors such as manufacturing and chemicals. The rapid pivot away from Russian energy imports following the invasion led to skyrocketing energy costs, supply chain disruptions, and factory closures across Europe. Many Europeans, especially during the winter months, experienced energy rationing and rising living costs.

The lessons from this conflict will shape international relations for decades to come. It's up to leaders and citizens alike to ensure these lessons lead to a more stable and peaceful world, rather than a continuation of the cycle of mistrust and conflict.

On June 24, 2024, leaders of the Group of Seven (G-7) wealthy democracies agreed to engineer a $50 billion loan package for Ukraine, funded by interest accrued on approximately $300 billion in frozen Russian central bank assets held in Western jurisdictions. This represents an unprecedented financial maneuver, raising significant legal and geopolitical concerns. President Vladimir Putin called the G7 deal "theft" and warned of long-term consequences. Critics argue that the move risks undermining global confidence in Western financial institutions, particularly among countries outside the G-7 bloc.

If the G-7 leaders have set a precedent of repurposing sovereign assets, could it create a chilling effect for other countries—especially emerging economies—holding reserves in G-7 banks? Observers warn of potential capital flight toward non-Western financial hubs or alternative currencies such as the Chinese yuan or BRICS-backed instruments.

NATO's strategic pivot toward the Indo-Pacific and recent rhetoric targeting China's role in supporting Russia has heightened tensions. On July 10, 2024, NATO formally labeled China a "decisive enabler" of Russia's war in Ukraine, a significant diplomatic escalation. This suggests NATO's ambition to expand its focus into Asia, particularly via security cooperation with countries like

Japan, South Korea, and Australia. Many analysts view this as a risky entanglement that could spark broader conflict in the Pacific.

China remains the world's longest continuous civilization, with over 4,000 years of recorded history. The trauma of the "Century of Humiliation" (1842–1949), which included the Opium Wars, unequal treaties, and foreign occupation, has shaped national identity and policy. Modern Chinese leaders, especially under Xi Jinping, have vowed to prevent a repeat of this period, emphasizing national sovereignty, military modernization, and economic self-reliance.

For centuries, China was the leading power in science, technology, and trade, until dynastic stagnation and foreign intervention eroded its dominance. Today, China's population, technological progress, and military strength make it a formidable global actor. In the event of a direct military confrontation, particularly one involving nuclear powers, mutual destruction would be nearly inevitable—making de-escalation a critical global priority.

Shifting Political Landscapes in 2024–2025

The second half of 2024 saw sweeping political change within the G-7 countries. Germany's ruling Social Democratic Party (SPD), led by Chancellor Olaf Scholz, received only 14% in the European Parliament elections on June 9, signaling waning public confidence. In Italy, Prime Minister Giorgia Meloni's Brothers of Italy party dominated with 28%, consolidating her far-right coalition's power.

In the UK, Keir Starmer became the seventh prime minister in 17 years after leading the Labour Party to victory on July 5. France's President Emmanuel Macron suffered a major defeat in the July 7 legislative elections, resulting in the resignation of his government on July 16. Though Macron remains president, his political influence has significantly weakened.

On July 21, U.S. President Joe Biden withdrew from the 2024 presidential race, citing health concerns and political pressure. Vice President Kamala Harris was named the Democratic nominee, assuming Biden's campaign funds despite not winning a single primary.

In Japan, Shigeru Ishiba replaced Fumio Kishida as Prime Minister on October 1, making him the ninth PM since 2008, reflecting ongoing political instability in the country. In Canada, following repeated by-election losses and growing dissent within his own party, Prime Minister Justin Trudeau resigned on January 7, 2025.

Mark Joseph Carney, a former central banker and climate finance advocate, became the new leader of Canada's Liberal Party on March 9 and assumed office as prime minister on March 14, 2025. Carney previously served as Governor of the Bank of Canada (2008–2013) and Governor of the Bank of England (2013–2020), making him the only person to have led both institutions.

Economic Upheaval: Trump's Return and Global Fallout

Donald Trump won the U.S. presidential election on November 5, 2024, in a landslide, campaigning on economic nationalism and trade protectionism. On April 2, 2025, shortly after markets closed, President Trump announced a sweeping global tariff plan. The Dow Jones Industrial Average (DJIA) closed at 42,215.24 and the NASDAQ at 17,596.71. Within three trading days, the

DJIA plummeted to 37,965.60 (-10.07%) and the NASDAQ fell to 15,603.26 (-11.33%), wiping out over $10 trillion in global stock market capitalization.

Tech stocks were among the hardest hit: Dell Technologies dropped 19%, HP fell 15%, and Western Digital plunged 18%. Analysts described the tariff plan as an economic act of aggression that triggered global panic. In response, Canadian Prime Minister Mark Carney declared an end to Canada's post-WWII economic reliance on the U.S., stating:

"The system of global trade anchored on the United States… is over. Our old relationship of steadily deepening integration with the United States is over."

EU Commission President Ursula von der Leyen warned that Trump's tariffs would cause consumer prices to rise, disrupt supply chains, and spike inflation, especially in critical sectors like pharmaceuticals and agriculture.

China, facing a proposed 54% U.S. tariff on its exports, retaliated with a 34% levy on American goods. Trump threatened to nearly double U.S. tariffs in response. According to the BBC:

"Beijing's Commerce Ministry vowed to 'fight till the end,' calling Washington's move blackmail."

U.S. firms importing from China could soon face up to 104% total import taxes, making business unsustainable for many companies reliant on Chinese manufacturing. Economists warn this tit-for-tat cycle risks igniting a full-scale trade war between the world's two largest economies, with long-lasting consequences for global inflation, recession risks, and supply chain stability.

The BRICS summit was held in Kazan, Russia, on October 24, 2024. According to Reuters, here's a brief summary of the outcome of the BRICS summit:

- *XI AND MODI:* Chinese President Xi Jinping and Indian Prime Minister Narendra Modi met just two days after New Delhi announced that it had reached a deal with Beijing to resolve a four-year military stand-off on their Himalayan frontier.
- *LOTS OF LEADERS ATTENDED:* Putin, who wanted to show that the West's attempt to isolate Russia over the Ukraine war has failed, was able to attract major leaders such as Xi and Modi and nearly 20 others, including Turkey's Tayyip Erdogan. U.N. Secretary-General Antonio Guterres also attended.
- *BRICS WAITING LIST:* Putin said more than 30 countries had expressed a desire to join the BRICS, though there was little immediate clarity on how the expansion would work.
- *UKRAINE WAR:* BRICS leaders did raise the Ukraine war with Putin in different formats, but there was no sign that anything specific would be done to end the conflict.
- *BRICS MONEY:* BRICS predicted its influence would grow and outlined common projects ranging from a grain exchange to a cross-border payments system.
- *MIDDLE EAST:* Putin told BRICS leaders that the Middle East was on the brink of a full-scale war after a sharp rise in tension between Israel and Iran.

In December 2024, Assad's regime in Syria fell to terrorists. This was a surprise to most countries. Syria is now ruled under Sharia law. As we navigate complex issues and political power shifts, including de-dollarization,

I believe faith will continue to play a vital role in shaping human societies and individual lives. Whether through traditional religious institutions or new expressions of spirituality, people will continue to seek meaning, purpose, and connection to the divine.

In conclusion, while the future of faith may be uncertain in many ways, the human quest for spiritual truth and meaning remains constant. As we face the challenges of our time, may we approach the future with hope, wisdom, and a commitment to living out our deepest values and beliefs! Regardless of one's specific faith tradition, the call to love, compassion, and justice can guide us as we work towards a better future for all of humanity.

Chapter 41
My Perspective on the Origin of Life and Health Scare

The origin of life, according to science, began with the formation of simple inorganic molecules into organic compounds—amino acids, sugars, and lipids—eventually leading to the complex macromolecules that make up DNA and cells. While scientists have recreated basic building blocks in laboratory conditions, such as through the Miller-Urey experiment, the precise mechanisms that formed the first living cells remain unknown. Even the simplest single-cell organisms like bacteria contain a genetic program far too complex to be the product of random molecular collisions. To date, no laboratory or theory has successfully explained how functional DNA coding sequences arose spontaneously.

All computer programs are designed by programmers using binary code—combinations of "0" and "1." Similarly, the genetic code of all living beings is written using four chemical bases: Adenine (A), Cytosine (C), Guanine (G), and Thymine (T). These genetic "letters" form a language far more intricate than binary. DNA provides the instructions for each cell, telling it where it fits within the body and what functions to perform—whether to become a liver cell, a neuron, or part of a limb. The probability of this happening by pure chance is infinitesimally small.

Humans have souls, which are separate from their bodies. I know this for sure because I saw my mother's soul leave her body about five minutes after she stopped breathing. It was a moment of deep spiritual clarity, beyond scientific explanation. Evolution did not create souls. I know that God is real because I experienced God's spirit when I least expected it. No hallucination, dream, or emotion compares to the transformative certainty that comes with experiencing God's presence.

The authors of the Bible did not have a full scientific understanding when they wrote in the Gospel of John (Chapter 1:1–4):

"In the beginning was the Word, and the Word was with God, and the Word was God. The same was in the beginning with God. All things were made by Him, and without Him was not anything made that was made. In Him was life, and the life was the light of men."

We now understand that the "Word" can be metaphorically linked to the genetic code (DNA)—the foundational language of life. All organisms are built from this intricate code. Such complexity and precision reflect design, not random accident. I have examined all possible scientific origins of life, and the only consistent truth is that God is the Creator. No matter how many billions of years you allow, the DNA code will not randomly generate itself— because information always comes from intelligence.

God is real, and people need to know this truth. They need the assurance that God loves them deeply and is available to guide, comfort, and help those who seek Him in prayer and supplication.

The Federal Reserve and Global Control

The Federal Reserve has enabled wars that killed over 100 million people and injured many more. Its influence on monetary policy and military-industrial funding has shaped world conflicts from WWI to Iraq. Beyond war, it has fostered a system of global debt, trapping entire nations and populations under compounding interest systems. The Fed is the most lethal financial weapon ever created—and it must be abolished. A decentralized, transparent financial system could foster true economic justice. When people refuse to be slaves to fiat currencies, we will begin to move toward a more enlightened world.

An Age of Converging Crises

The world is overwhelmed by simultaneous challenges: financial instability, wars, pandemics, climate change, artificial intelligence, misinformation, social media overload, and civil unrest. Take time to refocus your mind. Many advanced civilizations—like the Maya, Khmer, and Akkadian—collapsed due to climate shifts that led to water and food scarcity, triggering wars and migration crises. We must act before it's too late.

AI can be incredibly helpful to humanity. It can also become an existential threat. Unchecked AI development could outpace ethical and spiritual guidance. We must learn to discern—for ourselves and for future generations.

Cryptocurrencies, while innovative in design, are increasingly showing signs of speculative instability and unsustainable energy use. Many will likely be revealed as Ponzi schemes in time. They also consume immense amounts of electricity and water, harming the environment. Bitcoin alone consumes more energy annually than some countries.

When no one can buy or sell without digital currency and QR codes, we risk sacrificing our freedoms, privacy, and even lives. Digital control can become totalitarian if misused. Lifelong learning will be essential to thrive in the

21st century and beyond. Entrepreneurship, creativity, and global collaboration will shape the new economy.

When your body dies, your soul will live on eternally. Don't be short-sighted. Plan your horizon beyond this world. Prophecies from various traditions—Christian, Jewish, Islamic, Indigenous, and others—foretell a dramatic transformation of the world. The end of this age may be nearer than most expect. Share scripture and love one another as God loves us. Worship only God.

Reflections on Health, Food, and Gratitude

Today in the U.S., people have to pay far more for organic food. Some organic produce costs 3 to 4 times more than non-organic food. Most Americans cannot afford to eat organic all the time. Ironically, in the poorest country— Burma (now Myanmar)—where I was born, all the food I ate growing up was organic by default. We didn't have pesticides or GMOs, just clean, natural produce and fresh meals.

When I came to the United States at age 14, I was about 5'4" and weighed 110 pounds. I was fit and healthy. But within a year of arriving in San Francisco, I was almost obese. I was eating processed and fried foods like Kentucky Fried Chicken (before it was called KFC), curry chicken wings, donuts, fried rice, and drinking Coca-Cola. I gained at least 50 pounds in one year. In college, I worked to bring my weight down to 129 pounds. This taught me that lifestyle and food choices make a profound difference. Choose wisely.

During the pandemic, I checked my mother's blood pressure and blood sugar almost every day, administering medications three times daily. I got only 3 to 4 hours of interrupted sleep most nights, as she often needed help at night. I neglected my own health during that time. Before the pandemic, I had no health issues. But one evening in 2022, after washing the dinner dishes, my left fingers felt tingly. I was surprised—and I immediately knew something wasn't right.

I called my sister Grace to inform her of the situation. Then I lay down and felt numbness spreading across the left side of my body. Within 30 minutes, I was in the ICU at Sutter Hospital. The next day, I underwent open-heart surgery. Without Grace and her eldest son, Joshua, I may not have survived. Their quick action saved my life.

I thank God and the medical professionals at Sutter Medical for preserving my life. I am still here today, writing this book, and I remain grateful every single day.

For my soul, writing this book is part of my journey and mission on Earth. You are here on Earth at this time for a reason. What does your soul want to do? What is your mission? What are your objectives? Be who you are meant to be.

Chapter 42
Blessings, Well Wishes, and
One Last Thing

Appreciate what you have, cherish your loved ones, and embrace the journey of lifelong learning. May the God who formed you in love—the One who knows you more deeply than you know yourself—bless you and keep you always. May He make His face shine upon you and fill you with His peace, joy, and strength. May He grant you the courage to be brave, the wisdom to make good choices, and the heart to love others as He loves you.

Moreover, may you always remember that you are a masterpiece—created by the God who hung the stars in the sky and painted the colors of the rainbow. You were fearfully and wonderfully made (Psalm 139:14), and there is no one else exactly like you. May you never doubt how special and precious you are to Him, or how much He delights in every part of who you are.

So go out into the world with a smile on your face, knowing that you are loved beyond measure and that your life is in the hands of the God who created you for a divine purpose. Keep running the race of faith with perseverance and passion, as encouraged in Hebrews 12:1–2, knowing that Jesus is right beside you, cheering you on every step of the way.

Furthermore, may you be a light in the darkness, a voice for the voiceless, and a friend to the friendless. May you use your unique gifts and talents to make the world a brighter, better, and more beautiful place. Your life matters. Your story matters. Your kindness and faith have the power to impact generations. May you always know that your life has meaning, value, and the power to change lives.

May the grace of our Lord Jesus Christ, the love of God, and the fellowship of the Holy Spirit be with you always (2 Corinthians 13:14).

Jesus said that nothing should distract us from preaching the gospel to all nations. In Matthew 28:19, He gave the Great Commission: "Go therefore and make disciples of all nations." Today, with the Internet and smartphones, billions of people have access to the Bible online—many in their native languages. With this book, it is my hope and prayer that more people will come to know Jesus Christ as their Lord and Savior.

I know that this book may be banned in some countries precisely because of this final paragraph. But I also believe that God can open doors that no one can shut (Revelation 3:8), and that He can soften hearts, even in the most unlikely places. I trust in God's timing, His wisdom, and His power to reach people wherever they are.

In life there are many mountains to climb. Whether you are climbing for fun, justice or other reasons, choose wisely. To God be all the glory.

Glossary

AI	Artificial Intelligence is the simulation of human intelligence processes by machines.
AIPAC	American Israel Public Affairs Committee, a pro-Israel lobbying group.
Amtrak	The National Railroad Passenger Corporation is a passenger railroad service that operates in the United States.
Angel Investor	An individual who provides capital for a business start-up, usually in exchange for convertible debt or ownership equity.
Anthropogenic	Caused or produced by humans
Anunnaki	Deities in ancient Mesopotamian cultures.
Apostate Church	A religious institution that outwardly appears Christian but has departed from true faith.
Artificial General Intelligence (AGI)	AI that matches or exceeds human intelligence across all domains.
Artificial Intelligence (AI)	The simulation of human intelligence processes by machines, especially computer systems.
Artificial Superintelligence (ASI)	AI that vastly surpasses human cognitive abilities in virtually all domains.
Big Bang	The theoretical starting point of the universe involves a massive expansion of space and time from an initial singularity.
Biodiversity	The variety of life in a particular habitat or ecosystem

Biotech	Short for biotechnology, the use of biological processes, organisms, or systems to manufacture products intended to improve the quality of human life.
Biotechnology	The use of biological processes, organisms, or systems to manufacture products intended to improve the quality of human life.
Blockchain	A decentralized, distributed ledger technology that records transactions across many computers.
BRICS	An economic group initially consisting of Brazil, Russia, India, China, and South Africa.
Carbon pricing	A method of charging for carbon emissions to reduce their production
Central Bank Digital Currency (CBDC)	A digital form of a country's fiat currency issued and regulated by the central bank.
CRISPR-Cas9	A unique technology that enables geneticists and medical researchers to edit parts of the genome by removing, adding, or altering sections of the DNA sequence. It is currently the simplest, most versatile, and precise method of genetic manipulation to modify gene function.
CRISPR	A gene-editing technology that allows for precise modifications of DNA sequences.
Cuneiform	An ancient writing system using wedge-shaped marks
Digital Currency	Money that exists only in electronic form and is managed through computer systems.
Dim sum	It is a large range of small Chinese dishes that are traditionally enjoyed

Simon Chin

in restaurants for brunch.

DNA (deoxyribonucleic acid) The molecule that carries the genetic instructions for the development, functioning, growth, and reproduction of all known organisms.

Edwards Air Force Base A United States Air Force installation located in southern California, known for its history of aeronautical research and development.

Epigenetics The study of changes in organisms caused by modification of gene expression rather than alteration of the genetic code itself.

ESL English as a Second Language, a program for non-native English speakers.

Ethics Moral principles that govern a person's behavior or the conducting of an activity.

Evolution The process by which species change over time through the inheritance of genetic variations across generations.

Exoplanet A planet that orbits a star other than the Sun.

Federal Reserve The central banking system of the United States

Fiat currency Money that is not backed by a physical commodity like gold

Four Horsemen of the Apocalypse Symbolic figures in the Book of Revelation represent conquest, war, famine, and death.

Genomics The study of all of a person's genes (the genome), including interactions of those genes with each other and

322

with the person's environment.

Glass-Steagall Act

A 1933 law that separated commercial and investment banking activities

Göbekli Tepe

An ancient archaeological site in Turkiye dated around 8000 BCE.

Global Financial Crisis (GFC) of 2007-2009

The most severe worldwide economic crisis since the Great Depression. Also known as the Great Recession

Great Depression (1929–1939)

The most severe global economic downturn that affected many countries across the world.

Great Recession

See Global Financial Crisis (GFC) of 2007-2009

Hamas

A Palestinian Sunni Islamist political and military organization

Homo sapiens

The scientific name for modern humans

Human Genome Project

An international scientific research project aimed at determining the sequence of the human genome and identifying and mapping all human genes.

Ignatian Center

A center at Santa Clara University that promotes and enhances the distinctively Jesuit Catholic tradition of education.

Indulgences

In Catholic tradition, a way to reduce the amount of punishment one has to undergo for sins.

Initial Public Offering (IPO)

The process of offering shares of a private corporation to the public in a new stock issuance.

Intellectual Property (IP)

Creations of the mind, such as inventions, literary and artistic works, designs, symbols, names, and images, are used in commerce.

Internet of Things (IoT)
The network of physical objects is embedded with sensors, software, and other technologies for the purpose of connecting and exchanging data with other devices and systems over the Internet.

ISO Class 1 cleanroom
A controlled environment with a low level of pollutants, used in manufacturing or scientific research that requires a high level of air purity.

Lifelong Learning
The ongoing, voluntary, and self-motivated pursuit of knowledge for personal or professional reasons.

Light pollution
Excessive or inappropriate artificial light that affects the ability to see the night sky.

Machine Learning
A subset of AI that enables systems to learn and improve from experience without explicit programming.

Mansa Musa
The 14th-century emperor of the Mali Empire was considered the wealthiest person in history.

Mark of the Beast
A prophesied mark is required for buying and selling in the end times.

Melatonin
A hormone that regulates the sleep-wake cycle, with its production increasing in the evening and decreasing in the morning.

Metabolites
Small molecules that are intermediates and products of metabolism.

Microbiome
The collection of all microbes, such as bacteria, fungi, and viruses, that live on and in our bodies.

Middle Passage	The stage of the Atlantic slave trade in which millions of enslaved Africans were transported to the Americas.
Military-industrial complex	The relationship between a country's military, defense industry, and political leadership
NATO	North Atlantic Treaty Organization, a military alliance between North American and European countries
Natural selection	The process by which organisms with advantageous traits are more likely to survive and reproduce, leading to changes in species over time.
Neural Networks	Computing systems inspired by biological neural networks are capable of machine learning and pattern recognition.
Paleogenomics	The study of ancient DNA
Patent	A government authority or license conferring a right or title for a set period, especially the sole right to exclude others from making, using, or selling an invention.
Pharmacogenomics	The study of how genes affect a person's response to drugs.
Photomask	A plate with holes or transparencies that allow light to shine through in a defined pattern is used in photolithography. A photomask set of 70 to 80 distinct layers is required to create an advanced semiconductor with features in the nanometers.
Protestant Reformation	A 16th-century movement for the reform of the Roman Catholic Church that led to the establishment of Protestant churches.

QR Code	A type of matrix barcode that can store various types of data and is readable by smartphones.
Quantitative easing	A monetary policy where a central bank purchases securities to increase the money supply
Quantum mechanics	The branch of physics that describes the behavior of matter and energy at the atomic and subatomic levels.
Rangoon (Yangon)	It is the former capital city of Burma (Myanmar).
Rapture	The belief is that believers will be taken from Earth to meet Christ in the air before or during the tribulation.
Resilience	The capacity to recover quickly from difficulties; toughness.
Rotary International	A global non-profit service organization that brings together business and professional leaders to provide humanitarian service and advance goodwill and peace around the world.
Semiconductor	A material with electrical conductivity between that of a conductor and an insulator.
Singularity	A point in space-time at which gravitational forces cause matter to have infinite density and infinitesimal volume and space and time to become infinitely distorted.
Soul	The spiritual part of a human being is regarded as immortal.
Spiritual walkabout	A journey of spiritual discovery and growth.
STEM	Science, Technology, Engineering,

and Mathematics.

Sustainable
Able to be maintained at a certain rate or level without depleting resources

Theory of relativity
Albert Einstein's theory that describes gravity as a warping of space-time caused by the presence of mass and energy.

Thingyan
The Burmese New Year festival is celebrated in mid-April.

Trail of Tears
A series of forced relocations of Native American tribes in the United States in the 1830s.

Tribulation
A period of great suffering and distress is predicted in biblical end-time prophecies.

UC Berkeley
University of California, Berkeley, a prestigious public research university

UNESCO
United Nations Educational, Scientific and Cultural Organization.

US Borax
A mining company that supplies nearly half the world's refined borates, minerals essential to life and modern living.

USDA
United States Department of Agriculture, a federal agency responsible for developing and executing federal laws related to farming, forestry, rural economic development, and food.

Vatican City
An independent city-state and the headquarters of the Roman Catholic Church.

Venture Capital (VC)
A form of private equity financing that is provided by venture capital firms or funds to startups, early-stage, and emerging companies that

Simon Chin

have been deemed to have high growth potential.

YMCA

The Young Men's Christian Association is a worldwide organization with a mission to put Christian principles into practice through programs that build a healthy spirit, mind, and body for all.

www.ingramcontent.com/pod-product-compliance
Lightning Source LLC
Chambersburg PA
CBHW060125130626
46556CB00006B/2230